Guidance Note 7

Special Locations

18th IET Wiring Regulations BS 7671:2018

Published by The Institution of Engineering and Technology, London, United Kingdom

The Institution of Engineering and Technology is registered as a Charity in England & Wales (no. 211014) and Scotland (no. SCO38698).

 The Institution of Engineering and Technology is the institution formed by the joining together of the IEE (The Institution of Electrical Engineers) and the IIE (The Institution of Incorporated Engineers).

First published 1998 (0 85296 601 6)

Second edition (incorporating Amendment No. 1 to BS 7671:2001) 2003 (0 85296 995 3)
Reprinted (incorporating Amendment No. 2 to BS 7671:2001) 2004
Third edition (incorporating BS 7671:2008) 2009 (978-0-86341-861-7)
Fourth edition (incorporating Amendment No 1 to BS 7671:2008) 2011 (978-1-84919-283-5)
Fifth edition (incorporating Amendment Nos. 2 and 3 to BS 7671:2008) 2015
(978-1-84919-891-3)
Reprinted October 2015
Sixth edition (incorporating BS 7671:2018) 2018 (978-1-78561-464-4)

Copies of this publication may be obtained from:
PO Box 96, Stevenage, SG1 2SD, UK
Tel: +44 (0)1438 767328
Email: sales@theiet.org
www.theiet.org/wiringbooks

ISBN 978-1-78561-464-4 (paperback)
ISBN 978-1-78561-465-1 (electronic)

Typeset in the UK by the Institution of Engineering and Technology, Stevenage

Printed in the UK by Sterling Press Ltd, Kettering

Contents

Cooperating organisations

The Institution of Engineering and Technology acknowledges the invaluable contribution made by the following individuals in the preparation of this Guidance Note:

Institution of Engineering and Technology

M. Al-Rufaie BSc CEng MIET FIHEEM
M. Coles BEng(Hons) MIET
M. Cotterell Co-convenor IEC TC 82 WG3 and member of GEL 82
H.R. Lovegrove IEng FIET
Eur Ing L. Markwell MSc BSc(Hons) CEng MIET MCIBSE LCGI
D. Roberts BSc CEng MIMechE and Chairman of CPW/4
G.G. Willard DipEE CEng FIET JP

We would like to thank the following organisations for their continued support:

AMEDA
AXREM
BEAMA Installation
Bender UK
Brandon Medical Company
British Cables Association
BSI Committees
Cableflow International Limited
Certsure trading as NICEIC and Elecsa
ECA
SELECT
Electrical Safety First
The GAMBICA Association Ltd
Hager Engineering Ltd
Health and Safety Executive
Institute of Healthcare Engineering and Estate Management
Lighting Industry Association
Medical Device and Estates Safety Policy Branch - Safety Strategy Unit (Health Service Northern Ireland).
Megger Instruments Ltd
NAPIT
National Caravan Council Ltd
NHS Estates and Facilities Efficiency & Productivity Division (NHS improvement - Health Service England).

NHS National Services Scotland (Health Service Scotland).
NHS Wales Shared Services Partnership - Specialist Estates
(Health Service Wales)
Safety Assessment Federation (SAFed)
Siemens Healthcare Limited
Society of Electrical and Mechanical Engineers serving Local Government
Starkstrom UK
Underfloor Heating Manufacturers' Association

Guidance Note 7 revised, compiled and edited

G.D. Cronshaw CEng FIET and P.E. Donnachie BSc CEng FIET

Acknowledgements

References to British Standards, CENELEC Harmonization Documents and International Electrotechnical Commission (IEC) standards are made with the kind permission of BSI. Complete copies can be obtained by post from:

BSI Customer Services
389 Chiswick High Road
London W4 4AL
Tel: +44 (0)20 8996 9001
Fax: +44 (0)20 8996 7001
Email: cservices@bsigroup.com

BSI operates an export advisory service – Technical Help to Exporters

(+44 (0)20 8996 7111) – that can advise on the requirements of foreign laws and standards. The BSI also maintains stocks of international and foreign standards, with many English translations. Up-to-date information on BSI standards can be obtained from the BSI website: www.bsigroup.com

The illustrations within this publication were provided by Rod Farquhar Design: www.farquhardesign.co.uk

Cover design and illustration were created by The Page Design: www.thepagedesign.co.uk

Copies of Health and Safety Executive documents and approved codes of practice (ACOP) can be obtained from:

HSE Books
PO Box 1999
Sudbury, Suffolk CO10 2WA
Tel: +44 (0)1787 881165
Email: hsebooks@prolog.uk.com
Web: http://books.hse.gov.uk/hse/public/home.jsf

The HSE website is www.hse.gov.uk

Preface

This Guidance Note is part of a series issued by the Institution of Engineering and Technology to explain and enlarge upon the requirements in BS 7671:2018, the 18th Edition of the IET Wiring Regulations.

This Guidance Note does not necessarily ensure compliance with BS 7671. It is intended to explain some of the requirements of BS 7671, but readers should always consult BS 7671 to satisfy themselves of compliance.

The scope generally follows that of BS 7671; the relevant regulations and appendices are noted in the margin. Some Guidance Notes also contain material not included in BS 7671:2018 but which was included in earlier editions of the Wiring Regulations. All of the Guidance Notes contain references to other relevant sources of information.

Electrical installations in the United Kingdom that comply with BS 7671 are likely to satisfy Statutory Regulations such as the Electricity at Work Regulations 1989, however, this cannot be guaranteed. It is stressed that it is essential to establish which statutory and other regulations apply and to install accordingly. For example, an installation in premises subject to licensing may have requirements different from, or additional to, BS 7671 and these will take precedence.

Persons carrying out electrical work in many of these special locations and installations must meet the requirements of the relevant Building Regulations. In England, the Building Regulations 2010 apply (Approved Document Part P (2013) – *Electrical safety – Dwellings*). In Scotland, the requirements of the Building Regulations (Scotland) 2004 apply. In Wales, the Building Regulations 2010 apply (Approved Document Part P (2006) – *Electrical safety – Dwellings*). In Northern Ireland, the Building Regulations (Northern Ireland) 2000 apply.

Introduction

General

This Guidance Note has been revised to align with BS 7671:2018 *IET Wiring Regulations* 18th Edition. BS 7671:2018 includes the following additional section on special installations that was not included in the previous edition: Onshore units of electrical shore connections for inland navigation vessels (Section 730).

This Guidance Note has been revised where necessary to maintain alignment with the technical requirements of the latest IEC and European Committee for Electrotechnical Standardization (CENELEC) standards.

Note: A new IET *Code of Practice for Electric Vehicle Charging Equipment Installation* is available now.

It is to be noted that BS 7671 and this Guidance Note are concerned with the design, selection, erection, inspection and testing of electrical installations and that these documents may need to be supplemented by the requirements or recommendations of other British Standards.

110.1.3 Other standards of note are described in Regulation 110.1.3, including BS EN 60079 *Electrical apparatus for explosive gas atmospheres* (other than mines) and BS EN 50281 and BS EN 61241 *Electrical apparatus for use in the presence of combustible dust*.

Section 700 The particular requirements of Part 7 of BS 7671 supplement or modify the general requirements contained in the remainder of the standard. Thus, particular protective measures may not be allowed or supplementary measures may be required. However, it is important to remember that in the absence of any commentor requirement in Part 7, the relevant requirements of the rest of the Regulations must be applied.

The particular requirements within some Part 7 sections place a prohibition on the use of certain measures of protection, e.g. obstacles and placing out of reach.

International and European Standards

Preface Part 7 of BS 7671 technically aligns with the relevant CENELEC Harmonization Documents (HD) or draft Harmonization Documents (prHD). The preface to BS 7671 identifies the particular CENELEC Harmonization Documents current at the time of publication.

For those persons engaged in work outside the UK the Guidance Note advises whether BS 7671 is based on the European (HD) or the International (IEC) standard.

Contents

This Guidance Note discusses all the sections of Part 7 and includes chapters on special locations/installations not included in BS 7671 as follows:

▶ Chapter 13 Gardens, including supplies to outbuildings and hot tubs; and
▶ Chapter 15 Small-scale embedded generators (SSEG).

The guidance is based on published IEC standards and draft CENELEC proposals.

Exclusions

The guide does not consider those special installations or equipment where the requirements are specified in other British Standards, such as:

BS EN 60079   ▶   Electrical apparatus for explosive gas atmospheres;
BS EN 50281   ▶   Electrical apparatus for use in the presence of combustible dust;
BS EN 61241   ▶   Emergency lighting; and
BS 5839   ▶   Fire detection and fire alarm systems for buildings.
BS 5266
BS EN 1838

Locations containing a bath or shower 1

1.1 Introduction

Sect 701   There are no significant changes to Section 701 introduced by the 18th Edition.

1.2 Scope

701.1   As stated in Regulation 701.1, the particular requirements of Section 701 apply to
Sect 701   locations containing a fixed bath (bath tub, birthing pool) or shower and to the surrounding zones as described in the regulations. The requirements do not apply to emergency facilities in industrial areas or laboratories, on the presumption that they will only be used in an emergency. Where they are used with any regularity the requirements of the section would apply.

1.3 The risks

The following information is provided to give a better understanding of why particular requirements are necessary for bathrooms and other wet locations.

Persons in bathrooms are particularly at risk due to:

(a) lack of clothing, particularly footwear;
(b) presence of water, which reduces contact resistance;
(c) immersion in water, which reduces total body resistance;
(d) ready availability of earthed metalwork; and
(e) increased contact area.

1.3.1 Clothing

Clothing, particularly footwear such as shoes or boots, can greatly increase the total body resistance.

1.3.2 Body impedance

IEC publication DD IEC/TS 60479-1 *Effects of current on human beings and livestock* provides information on body impedance. Body impedance varies from person to person. The value of impedance depends on a number of factors, in particular on the current path, touch voltage, duration of current flow, frequency, degree of moisture of the skin, surface area of contact, pressure exerted and temperature. The IEC document provides information on different total body impedances hand-to-hand for small, medium and large surface areas of contact in dry, water-wet and saltwater-wet conditions. For bathrooms and showers, Table 1.1 shows the varying total body impedance for a current path hand-to-hand, for large surface areas (10000 mm^2) in contact with dry and water-wet conditions for different touch voltages.

Large surface areas of contact have the lowest impedances, with medium surface areas of contact (1000 mm²) and small surface areas of contact (100 mm²) having impedances of progressively higher magnitudes.

1.3.3 Immersion

Immersion of a body in bath water produces large areas of contact of water and, as Table 1.1 shows, this will reduce the total body impedance.

This reduction in body impedance coupled with a location that has earthed metalwork from pipes etc. makes a bathroom particularly hazardous and therefore requires special precautions to be taken.

▼ **Table 1.1** Total body impedances Z_T for a current path hand-to-hand AC 50/60 Hz, for large areas of contact in dry and water-wet conditions

Touch voltage	Values for the total body impedance Z_T (ohms) that are not exceeded for:					
	5 % of population		50 % of population		95 % of population	
(V)	Dry	Water-wet	Dry	Water-wet	Dry	Water-wet
25	1750	1175	3250	2175	6100	4100
50	1375	1100	2500	2000	4600	3675
75	1125	1025	2000	1825	3600	3275
100	990	975	1725	1675	3125	2950
125	900	900	1550	1550	2675	2675
150	850	850	1400	1400	2350	2350
175	825	825	1325	1325	2175	2175
200	800	800	1275	1275	2050	2050
225	775	775	1225	1225	1900	1900
400	700	700	950	950	1275	1275
500	625	625	850	850	1150	1150
700	575	575	775	775	1050	1050
1000	575	575	775	775	1050	1050
Asymptotic value = internal impedance	575	575	775	775	1050	1050

Notes:

(a) Some measurements indicate that the total body impedance for the current path hand-to-foot is somewhat lower than for a current path hand-to-hand (10 % to 30 %).

(b) For living persons the values of Z_T correspond to a duration of current flow of about 0.1 s. For longer durations Z_T values may decrease (about 10 % to 20 %) and after complete rupture of the skin Z_T approaches the internal body impedance Z_i.

(c) For the standard value of the voltage 230 V (network-system 3N ~ 230/400 V) it may be assumed that the values of the total body impedance are the same as for a touch voltage of 225 V.

(d) Values of Z_T are rounded to 25 Ω for dry and water-wet conditions.

1.4 Zones

701.32.2
701.32.3
701.32.4
Zones 0, 1 and 2 provide a very practical method of specifying requirements for the protection of equipment against the ingress of water and protection against electric shock, etc. in a specific and unambiguous way. Equipment is either in a zone or outside a zone and this can be determined by measurement.

▶ Figures 1.1a, b and g show the zones in relation to a bath tub.
▶ Figures 1.1c, d and h show the zones in relation to a shower.
▶ Figures 1.1e and f show the zones in relation to a shower without a basin from a fixed water outlet (shower head). Demountable shower heads are not considered and therefore do not alter the extent of the zone.

Horizontal or inclined ceilings, walls with or without windows, doors, floors and fixed partitions may be taken into account where these effectively limit the extent of locations containing a bath or shower as well as their zones.

Where a shower head is fixed to a wall or partition at a height greater than 2.25 m, the height of zone 1 extends to the same height as the shower head (see Regulation 701.32.3 (i)).

If there is a shower basin, there will be a zone 2 and its height will also extend to the same height as the shower head (see Regulation 701.32.4(i)).

The space under the bath is classed as zone 1 where it is accessible without the need for a tool. (Where the space is only accessible with a tool, it is considered to be outside the zones.) Current-using equipment permitted by Regulation 701.55 may be installed provided it is suitable for installation in zone 1 according to the manufacturer's instructions.

Examples of the zones for a bath and a shower are shown in Figures 1.1a-h.

▼ **Figure 1.1a**
Zones for a bath in elevation view

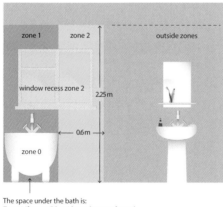

The space under the bath is:
Zone 1 if accessible without the use of a tool
Outside the zones if accessible only with the use of a tool

▼ **Figure 1.1b**
Zones for a bath in plan view

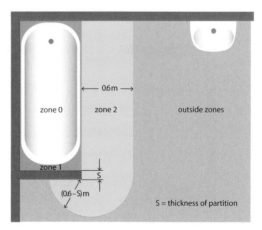

▼ **Figure 1.1c**
Zones for a shower in elevation view

▼ **Figure 1.1d**
Zones for a shower in plan view

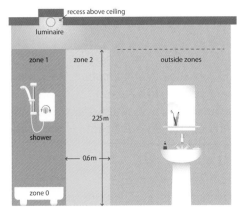

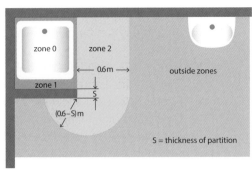

▼ **Figure 1.1e**
Zones for a shower, without basin

▼ **Figure 1.1f**
Zones for a shower, without basin, but with permanent fixed partition

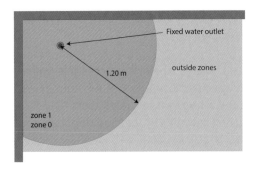

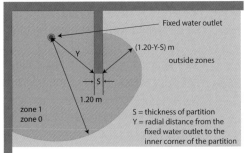

▼ **Figure 1.1g**
Zones for a bath in plan view adjacent to a cupboard

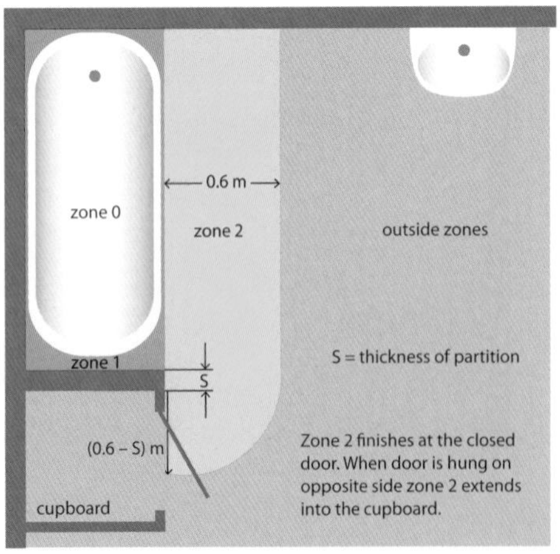

▼ **Figure 1.1h**
Zones for a shower in plan view adjacent to a cupboard

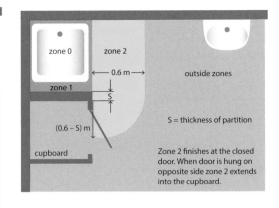

1.5 Protection against electric shock

1.5.1 Locations containing a bath or shower

Section 701 includes the following requirements:

701.411.3.3
415.1.1 ▶ Additional protection shall be provided for all low voltage circuits of the location, including circuits passing through zones 1 and/or 2 which do not serve the location, by the use of one or more RCDs having the characteristics specified in Regulation 415.1.1.

701.512.3
BS EN 61558-2-5 ▶ Except for SELV socket-outlets complying with Section 414 and shaver supply units complying with BS EN 61558-2-5, socket-outlets are prohibited within a distance of 3 m horizontally from the boundary of zone 1.

701.410.3.5 The protective measures of obstacles and placing out of reach (Section 417) shall not be used.

701.410.3.6 The protective measures of non-conducting location (Regulation 418.1) and earth-free local equipotential bonding (Regulation 418.2) shall not be used.

1.5.2 Supplementary protective equipotential bonding

701.415.2
411.3.1.2 Local supplementary protective equipotential bonding is required to be established in all areas within a room containing a bath tub or shower basin, by Regulation 701.415.2 (refer to Figure 1.2). **However, the last paragraph of the regulation allows this bonding to be omitted where the location containing a bath or shower is in a building with a protective equipotential bonding system in accordance with Regulation 411.3.1.2 and provided all of the following conditions are met:**

(**Note**: The labelling of the indents below differs from BS 7671.)

411.3.2 **(a)** All final circuits of the location comply with the requirements for automatic disconnection according to Regulation 411.3.2.
(b) All final circuits of the location have additional protection by means of an RCD in accordance with Regulation 701.411.3.3.
(c) All extraneous-conductive-parts of the location are effectively connected to the protective equipotential bonding according to Regulation 411.3.1.2.

This means that the designer/installer needs to establish that all extraneous-conductive-parts of the location are effectively connected to the protective equipotential bonding according to Regulation 411.3.1.2.

Note: The effectiveness of the connection of extraneous-conductive-parts in the location to the main earthing terminal may be assessed, where necessary, by the application of Regulation 415.2.2.

415.2.2 The resistance R (in ohms) between simultaneously accessible exposed-conductive-parts and extraneous-conductive-parts shall fulfil the following condition:

$R \leq 50\ V/I_a$ in AC systems
$R \leq 120\ V/I_a$ in DC systems

where:

I_a is the operating current in amperes of the protective device -
for RCDs, $I\Delta_n$.
for overcurrent devices, the current causing operation in 5 s.

▼ **Figure 1.2** Supplementary protective equipotential bonding in a bathroom – metal pipe installation

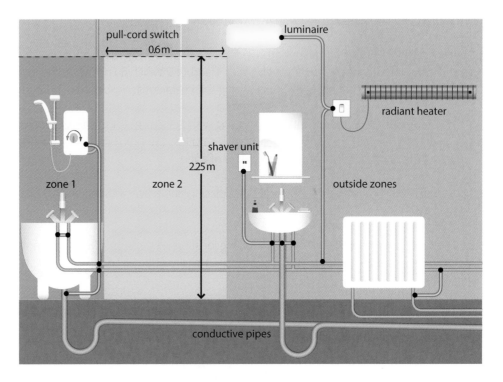

Notes:

(a) The protective conductors of each circuit supplying Class I and Class II equipment within the room must be supplementary bonded to all extraneous-conductive-parts within the room, including metal waste, water and central heating pipes, and metal baths and metal shower basins. Supplementary bonding of circuit protective conductors and extraneous-conductive-parts in the room does not require multiple connections.

(b) Circuit protective conductors may be used as supplementary bonding conductors.

(c) The space below the bath/shower is zone 1 if it can be accessed without the use of a tool, otherwise it is outside the zones.

In practice, the resistance between bonded extraneous-conductive-parts and exposed-conductive-parts should not exceed 0.05 ohm.

If supplementary bonding is carried out, the following points are worth noting:

701.415.2 ▶ Supplementary protective bonding does not necessarily have to be carried out within the bathroom itself but may be carried out in close proximity, such as under the floorboards, above the ceiling, or in an adjacent airing cupboard.

▶ The requirement to supplementary bond to the protective conductors of circuits supplying both Class I and Class II equipment is necessary in case, during the life of the installation, the user changes a Class II item of equipment for Class I.

1.5.3 Supplementary protective equipotential bonding of plastic pipe installations

Supplementary bonding is not required to metallic parts supplied by plastic pipes such as metal hot and cold water taps supplied from plastic pipes, or to a metal bath not connected to extraneous-conductive-parts such as structural steelwork and where the hot and cold water pipes and the waste are plastic (refer to Figure 1.3).

Supplementary bonding is also not required to short lengths of metal pipe that are often installed for cosmetic reasons when the basic plumbing system is plastic.

▼ **Figure 1.3** Supplementary protective equipotential bonding in a bathroom – plastic pipe installation

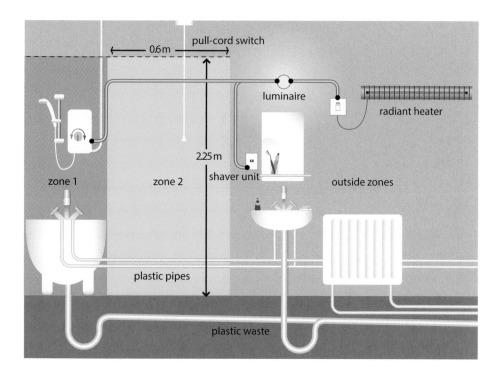

Notes:

(a) The protective conductors of all power and lighting points within the room must be supplementary bonded. The bonding connection may be to the earth terminal of a switch or an accessory supplying equipment. Supplementary bonding of circuit protective conductors and extraneous-conductive-parts in the zones does not require multiple connections.

(b) Circuit protective conductors may be used as supplementary protective bonding conductors.

(c) The space below the bath/shower is zone 1 if it can be accessed without the use of a tool, otherwise it is outside the zones.

1.6 Switchgear and controlgear

701.512.3 Switches and controls, other than those that are incorporated in fixed current-using equipment suitable for use in that zone or insulating pull cords of cord operated switches, are not allowed in zones 0, 1 or 2. This means that switches are allowed on showers and fans if the IP rating of the equipment, including that of the switch, is appropriate for use in the particular zone.

Exceptionally, switches of SELV circuits supplied at a nominal voltage not exceeding 12 V AC rms or 30 V ripple-free DC with the safety source installed outside zones 0, 1 and 2 may be installed in zone 1, or SELV up to 50 V AC rms or 120 V DC in zone 2. Also, shaver supply units complying with BS EN 61558-2-5 may be installed in zone 2.

Except for SELV socket-outlets complying with Section 414 and shaver supply units complying with BS EN 61558-2-5, socket-outlets are prohibited within a distance of 3 m horizontally from the boundary of zone 1 (refer to Figure 1.4).

▼ **Figure 1.4** Installation of a 230 V socket-outlet in a bathroom

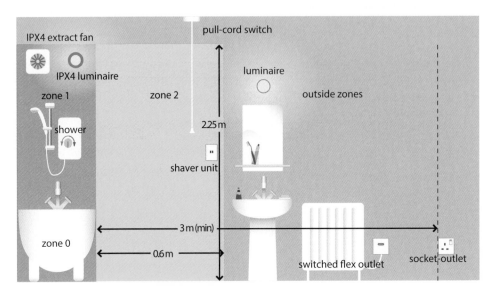

The requirements for switchgear, controlgear and accessories in locations containing a bath or shower are summarized in Table 1.2.

1.6.1 Telephones

Telephones and their sockets should be installed outside zones 0, 1 and 2.

1.7 Current-using equipment

701.55 Fixed and permanently connected current-using equipment may be installed in zones 1, 2 and outside the zones, but there are specific requirements for degrees of protection (see Table 1.2 for a summary of the requirements). Equipment having a rated voltage of 230 V may be installed in the above zones provided it has the appropriate IP rating and is suitable for use in the zone. This includes equipment suitable for use in the zone incorporating switches and controls.

Part 2 Current-using equipment is equipment that consumes current rather than simply transmits it or switches it. Examples include appliances, luminaires, fans and heaters.

Extractor fans

A suitable 230 V extractor fan may be installed in zone 1 or 2, or outside the zones.

If an extractor fan is installed in zone 1 or 2 it must be protected against the ingress of moisture to at least IPX4. An extractor fan supplied from a lighting circuit for a bathroom without a window should have its own means of isolation, as otherwise replacement or maintenance of the fan would have to be carried out in the dark. An isolation switch for a fan with an overrun facility will need to be triple-pole (switch wire, line and neutral), and must be installed outside the zones.

Note: Refer to the manufacturer for overcurrent protection.

▼ **Table 1.2** Summary of requirements for equipment (current-using and accessories) in locations containing a bath or shower

Zone*	Requirements for equipment in the zones		
	Minimum degree of protection	**Current-using equipment, e.g. appliance or luminaire**	**Switchgear, controlgear and accessories**
0	IPX7	Only suitable 12 V AC rms or 30 V ripple-free DC SELV fixed and permanently connected equipment that cannot be located elsewhere, the safety source being installed outside the zones.	None allowed.
1	IPX4 (Electrical equipment exposed to water jets, e.g. for cleaning purposes, shall have a degree of protection of at least IPX5)	The following fixed and permanently connected current-using equipment if suitable for installation in zone 1 according to the manufacturer's instructions: • whirlpool units (e.g. hot tubs and whirlpool baths); • electric showers; • shower pumps; • equipment protected by SELV or PELV at a nominal voltage not exceeding 25 V AC rms or 60 V ripple-free DC, the safety source being installed outside zones 0, 1 and 2; • ventilation equipment; • towel rails; • water heating appliances; and • luminaires.	Only switches of SELV circuits supplied at a nominal voltage not exceeding 12 V AC rms or 30 V ripple-free DC shall be installed, the safety source being outside zones 0, 1 and 2.

Zone*	Requirements for equipment in the zones		
	Minimum degree of protection	Current-using equipment, e.g. appliance or luminaire	Switchgear, controlgear and accessories
2	IPX4 (Electrical equipment exposed to water jets, e.g. for cleaning purposes, shall have a degree of protection of at least IPX5)	Current-using equipment if suitable.	Switchgear, accessories incorporating switches or socket-outlets shall not be installed with the exception of: switches and socket-outlets of SELV circuits, the safety source being installed outside zones 0, 1 and 2; and shaver supply units complying with BS EN 61558-2-5. Note: the IPX4 requirement does not apply to shaver supply units complying with BS EN 61558-2-5 installed in zone 2 and located where direct spray from showers is unlikely.
Outside zones	No additional requirement but subject to Regulation 512.2	No additional restrictions.	Except for SELV socket-outlets complying with Section 414 and shaver supply units complying with BS EN 61558-2-5, socket-outlets are prohibited within a distance of 3 m horizontally from the boundary of zone 1.

* See Figure 1.1 for zones.

1.8 Other equipment, e.g. home laundry equipment

Current-using equipment such as washing machines and tumble-dryers may be placed within zone 2 if it has a minimum degree of protection of IPX4 (if subject to water jets – at least IPX5 is required) and in accordance with the manufacturer's instructions permitted for such installation. Such equipment must be protected by a 30 mA RCD and supplied by means of a permanent connection unit located outside zone 2. Beyond 3 m horizontally from the boundary of zone 1 the equipment may be supplied by means of a plug and socket.

1.9 Electric floor heating systems

701.753 For electric floor heating systems, only heating cables according to relevant product standards or thin sheet flexible heating elements according to the relevant equipment standard may be erected provided that they have either a metal sheath or metal enclosure or a fine mesh metallic grid. The fine mesh metallic grid, metal sheath or metal enclosure must be connected to the protective conductor of the supply circuit. Compliance with the latter requirement is not required if the protective measure SELV is provided for the floor heating system. For electric floor heating systems, the protective measure 'protection by electrical separation' is not permitted.

Sect 753 BS 7671:2018 includes particular requirements for embedded heating systems (Section 753), for which guidance is provided in Chapter 12.

Swimming pools and other basins 2

2.1 Introduction

Sect 702 There are no significant changes to Section 702 introduced by the 18th Edition.

2.2 Scope

702.11 The particular requirements of Section 702 apply to the basins of swimming pools, the basins of fountains and the basins of paddling pools. The particular requirements also apply to the surrounding zones of these basins.

In these areas, in normal use, the risk of electric shock is increased by a reduction in body resistance and contact of the body with Earth potential. Swimming pools within the scope of an equipment standard are outside the scope of the regulations. Special requirements may be necessary for swimming pools for medical purposes. Except for areas especially designed as swimming pools, the requirements do not apply to natural waters, lakes in gravel pits, coastal areas and the like.

For private garden ponds and private garden fountains see Chapter 13.

2.3 The risks

The risk of electric shock is increased in swimming pools and their surrounding zones by the reduction in body resistance (see also bathrooms, section 1.3) and by good contact with Earth arising from wet partially clothed bodies. Equipment installed close to swimming pools and fountains is required to have appropriate degrees of protection against ingress of water.

2.4 Zones

702.32 The requirements for the classification of external influences are based on the dimensions of three zones (examples are given in Figures 702.1 to 702.4 of BS 7671:2018). For a swimming pool, zones 1 and 2 may be limited by fixed partitions having a minimum height of 2.5 m (see Figure 2.1).

▼ **Figure 2.1** Example of zone dimensions (plan) with fixed partitions of height at least 2.5 m

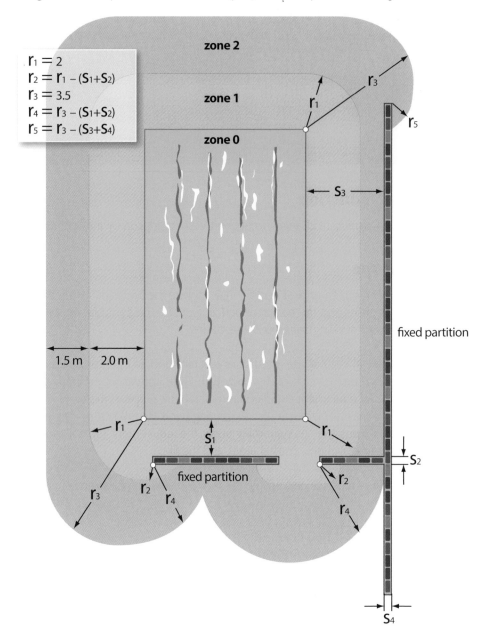

2.4.1   Zone 0

This zone is the interior of the basin of the swimming pool or fountain including any recesses in its walls or floors, basins for foot cleaning and water jets or waterfalls and the space below them.

2.4.2   Zone 1

This zone is limited by:

▶ zone 0;
▶ a vertical plane 2 m from the rim of the basin;
▶ the floor or surface expected to be occupied by persons; and
▶ the horizontal plane 2.5 m above the floor or the surface expected to be occupied by persons.

Where the swimming pool or fountain contains diving boards, springboards, starting blocks, chutes or other components expected to be used by persons, zone 1 comprises the zone limited by:

▶ a vertical plane situated 1.5 m from the periphery of the diving boards, springboards, starting blocks, chutes and other components such as accessible sculptures, viewing bays and decorative basins; and
▶ the horizontal plane 2.5 m above the highest surface expected to be occupied by persons.

2.4.3   Zone 2

This zone is limited by:

▶ the vertical plane external to zone 1 and a parallel plane 1.5 m from the former;
▶ the floor or surface expected to be occupied by persons; and
▶ the horizontal plane 2.5 m above the floor or surface expected to be occupied by persons.

There is no zone 2 for fountains (see Figure 2.2).

▼ **Figure 2.2** Example of determination of the zones of a fountain

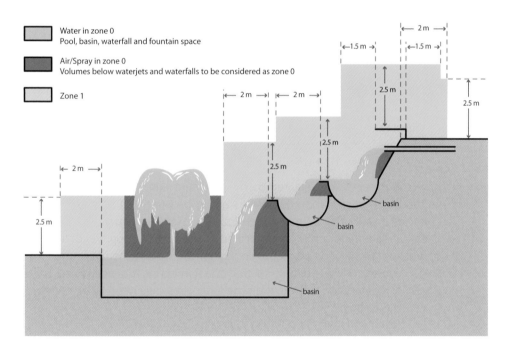

2.5   Protection for safety

702.410.3.5
702.410.3.6
Not surprisingly, the protective measures of obstacles and placing out of reach (Section 417) shall not be used. Also, the protective measures of non-conducting location (Regulation 418.1) and earth-free local equipotential bonding (Regulation 418.2) shall not be used.

702.415.2
702.52
Supplementary protective equipotential bonding is required between all extraneous-conductive-parts and the protective conductors of all exposed-conductive-parts in zones 0, 1 and 2, irrespective of whether they are simultaneously accessible, in accordance with Regulation 415.2. In zones 0, 1 and 2, any metallic sheath or metallic

covering of a wiring system (e.g. conduit), either surface-run or embedded in walls, floors etc. at a depth not exceeding 50 mm, must be connected to the supplementary bonding.

GN5 See Guidance Note 5 for test methods of determining an extraneous-conductive-part.

702.55.1 There is no particular requirement to install a metal grid in solid floors. However, where there is a metal grid, it must be connected to the local supplementary bonding. It is important to note that Section 753 'Heating cables and embedded heating systems' has requirements (such as a metal grid) where electric floor warming is installed.

The requirements for electric floor warming in swimming pool locations would be similar to the requirements for locations containing a bath or shower (see section 1.9).

2.6 Luminaires

2.6.1 Luminaires for swimming pools

702.410.3.4.1 In zones 0 and 1, only protection by SELV is permitted.

In zone 0, only protection by SELV at a nominal voltage not exceeding 12 V AC rms or 30 V ripple-free DC is permitted, the source for SELV being installed outside zones 0, 1 and 2.

In zone 1, only protection by SELV at a nominal voltage not exceeding 25 V AC rms or 60 V ripple-free DC is permitted, the source for SELV being installed outside zones 0, 1 and 2.

> **Note:** Equipment for use in the interior of basins that is only intended to be in operation when people are not inside zone 0 shall be supplied by a circuit protected by SELV, automatic disconnection of the supply (Section 411) using a 30 mA RCD, or electrical separation. See Regulation 702.410.3.4.1.

702.410.3.4.3 In zone 2, luminaires are required to be protected either by SELV, a 30 mA RCD or electrical separation. Where luminaires are to be protected by 30 mA RCDs, the high protective conductor currents often found in such equipment must be carefully considered. The luminaire manufacturer's advice should be sought with the objective of determining the standing protective conductor current and the maximum protective conductor current during starting so that the luminaires do not cause unwanted tripping. If RCD protection is used, then the luminaires may need to be on more than one circuit with separate RCDs.

Where there is no zone 2, section 2.8.1 describes the requirements to be fulfilled to enable 230 V luminaires to be installed in zone 1.

2.6.2 Underwater luminaires for swimming pools

702.55.2 A luminaire for use in the water or in contact with the water must be fixed and must comply with an appropriate part of BS EN 60598-2-18.

Underwater lighting located behind watertight portholes and serviced from behind must comply with the appropriate part of BS EN 60598 and be installed in such a way that no intentional or unintentional conductive connection between any exposed-conductive-part of the underwater luminaires and any conductive parts of the portholes can occur.

2.7 Socket-outlets

Socket-outlets must not be installed in zones 0 or 1, except for a swimming pool where it is not possible to install a socket-outlet outside zone 1. In this case, a socket-outlet that preferably has a non-conducting cover or coverplate may be installed in zone 1 provided that:

▶ it is installed at least 1.25 m from the border of zone 0 and at least 0.3 m above the floor; and

702.53 ▶ the supply to the socket-outlet is protected by:

- SELV (Section 414), at a nominal voltage not exceeding 25 V AC rms or 60 V ripple-free DC, with the source for SELV being installed outside zones 0 and 1; or
- automatic disconnection of supply (Section 411), using an RCD having the characteristics specified in Regulation 415.1.1; or
- electrical separation (Section 413), supplying only one item of current-using equipment, the source for electrical separation being installed outside zones 0 and 1.

Socket-outlets may be installed in zone 2 provided the supply circuit is protected by:

▶ SELV, with the source being installed outside zones 0, 1 and 2; or

▶ automatic disconnection of supply, using an RCD having the characteristics specified in Regulation 415.1.1; or

▶ electrical separation, with the source for electrical separation supplying only one socket-outlet and being installed outside zones 0, 1 and 2.

If using SELV or electrical separation to supply a socket-outlet in zone 2, the source can be installed in zone 2 provided the supply circuit is protected by an RCD having the characteristics specified in Regulation 415.1.1.

2.8 Current-using equipment of swimming pools

702.55.1 In zones 0 and 1, it is only permitted to install fixed current-using equipment specifically designed for use in a swimming pool, in accordance with the requirements of Regulations 702.55.2 and 702.55.4.

Equipment that is intended to be in operation only when people are outside zone 0 may be used in all zones provided it is supplied by a circuit protected according to Regulation 702.410.3.4.

It is permitted to install an electric heating unit embedded in the floor, provided it:

(a) is protected by SELV (Section 414), the source of SELV being installed outside zones 0, 1 and 2. However, it is permitted to install the source of SELV in zone 2 if its supply circuit is protected by an RCD having the characteristics specified in Regulation 415.1.1, or

(b) incorporates an earthed metallic sheath connected to the supplementary protective equipotential bonding specified in Regulation 702.415.2 and its supply circuit is additionally protected by an RCD having the characteristics specified in Regulation 415.1.1, or

(c) is covered by an embedded earthed metallic grid connected to the supplementary protective equipotential bonding specified in

Regulation 702.415.2 and its supply circuit is additionally protected by an RCD having the characteristics specified in Regulation 415.1.1.

Note: A socket-outlet and control device of equipment that is intended to be used in the interior of a swimming pool when not occupied should have a notice to warn the user that the equipment should only be used when the swimming pool is not occupied by persons.

2.8.1 Special requirements for the installation of electrical equipment in zone 1 of swimming pools and other basins

702.55.4 Fixed equipment designed for use in swimming pools and other basins (e.g. filtration systems, jet stream pumps) and supplied at low voltage is permitted in zone 1, subject to all the following requirements being met:

▶ The equipment must be located inside an insulating enclosure providing at least Class II or equivalent insulation and providing protection against mechanical impact of medium severity (AG2). This applies irrespective of the classification of the equipment.

▶ The equipment must only be accessible via a hatch (or a door) by means of a key or a tool. The opening of the hatch (or door) must disconnect all live conductors. The supply cable and the main disconnecting means should be installed in a way that provides protection of Class II or equivalent insulation.

▶ The supply circuit of the equipment must be protected by:

• SELV at a nominal voltage not exceeding 25 V AC rms or 60 V ripple-free DC, the source of SELV being installed outside zones 0, 1 and 2; or

• an RCD having the characteristics specified in Regulation 415.1.1; or

• electrical separation (Section 413), the source for electrical separation supplying a single fixed item of current-using equipment and being installed outside zones 0, 1 and 2.

A solution is included for the installation of 230 V luminaires for swimming pools where there is no zone 2.

Regulation 702.55.4 states that for swimming pools where there is no zone 2, lighting equipment supplied by other than a SELV source at 12 V AC rms or 30 V ripple-free DC may be installed in zone 1 on a wall or on a ceiling, provided that the following requirements are fulfilled:

(a) the circuit is protected by automatic disconnection of the supply and additional protection is provided by an RCD having the characteristics specified in Regulation 415.1.1; and

(b) the height from the floor is at least 2 m above the lower limit of zone 1.

In addition, every luminaire must have an enclosure providing Class II or equivalent insulation and providing protection against mechanical impact of medium severity.

2.9 Fire alarms and public address systems

Electrical equipment for safety systems such as fire alarms and public address may be required within the pool area. This equipment should be accessible for maintenance and preferably placed outside zones 0 and 1. Where it is not possible to locate equipment outside zone 1, it should be more than 1.25 m outside zone 0 and as high above floor level as is practicable, in order to keep the equipment dry. The equipment will need to be of an insulated construction and have IP coding as required by the zone. Local microphones must be SELV or connected via isolating transformers, and telephones, if required, should be the cordless type within the zones with a base unit installed outside the zones.

Reference should be made to the following standards:

▶ BS 5839 – *Fire detection and fire alarm systems for buildings;*
▶ BS EN 54 – *Fire detection and fire alarm systems;*
▶ BS 7445/BS EN 60849 – *Sound systems for emergency purposes;*
▶ BS 7827 – *Code of practice for designing, specifying, maintaining and operating emergency sound systems at sports venues.*

2.10   Fountains

2.10.1 General

The basins of fountains and their surroundings, unless persons are prevented from gaining access to them without the use of ladders or similar means, are treated as swimming pools.

702.410.3.4.2   In zones 0 and 1, one or more of the following protective measures must be used:

▶ SELV, the source for SELV being installed outside zones 0 and 1;
▶ automatic disconnection of supply using an RCD having the characteristics specified in Regulation 415.1.1; and
▶ electrical separation, the separation source supplying only one item of current-using equipment and being installed outside zones 0 and 1.

2.10.2 Electrical equipment of fountains

702.55.3   Electrical equipment in zones 0 or 1 must be provided with mechanical protection to medium severity (AG2), e.g. by use of mesh glass or by grids that can only be removed by the use of a tool.

Luminaires installed in zones 0 or 1 are required to be fixed and must comply with BS EN 60598-2-18.

An electric pump must comply with the requirements of BS EN 60335-2-41.

2.10.3 Additional requirements for the wiring of fountains

702.522.23   Cables supplying equipment in zone 0 should be installed outside the basin, i.e. in zone 1 or beyond where possible, and run to the equipment in the basin by the shortest practicable route.

For cables supplying equipment in zone 0, the designer should check with the supplier that the cable type is suitable for continuous immersion in water.

Cables supplying equipment in zone 1 should be selected, installed and provided with mechanical protection to medium severity (AG2) and the relevant submersion in water depth (AD8). Cable type H07RN8-F (BS EN 50525-2-21) is suitable up to a depth of 10 m of water. For depths greater than 10 m the cable manufacturer should be consulted.

2

Rooms and cabins containing sauna heaters

3

3.1 Introduction

Sect 703 There are no significant changes to Section 703 introduced by the 18th Edition.

3.2 Scope

703.1 The particular requirements of Section 703 apply to:

(a) sauna cabins erected on site, e.g. in a location or in a room; and

(b) the room where the sauna heater is, or the sauna heating appliances are installed. In this case, the whole room is considered as the sauna.

The requirements do not apply to prefabricated sauna cabins complying with a relevant equipment standard.

Where facilities such as showers, etc. are installed, the requirements of Section 701 will also apply.

3.3 The risks

There are two particular aspects of saunas that make them special locations:

(a) increased risk of electric shock because of extremely high humidity, lack of clothing, reduced skin resistance and large contact areas; and

(b) very high temperatures in certain zones.

3.3.1 Zones

703.32 The zones are temperature zones, dimensioned down from the ceiling, up from the
Fig 703 floor and around the sauna heater. This allows application of the zones whatever the size of the sauna cabin (see Fig 703 of BS 7671:2018).

3.4 Shock protection

703.410.3.5 The protective measures of obstacles and placing out of reach (Section 417) shall not
703.410.3.6 be used. Also, the protective measures of non-conducting location (Regulation 418.1) and earth-free local equipotential bonding (Regulation 418.2) shall not be used.

703.411.3.3 Additional protection must be provided for all circuits of the sauna, by means of one or more RCDs having the characteristics specified in Regulation 415.1.1. However, RCD protection need not be provided for the sauna heater itself unless such protection is recommended by the manufacturer, whose advice should be sought.

703.53 Protection against electric shock is also provided by not allowing any electrical equipment that is not part of the heating appliance or strictly necessary for the operation of the sauna, such as sauna thermostat, sauna cut-out and luminaires. Light switches must be placed outside the sauna room or cabin and socket-outlets should not be installed in the location containing the sauna heater. It is advisable not to install sockets near the cabin; the same criteria as for swimming pools should be adopted.

703.512.2 Only the sauna heater and equipment belonging to the sauna heater should be installed in zone 1.

703.1 A sauna is often part of a health or fitness suite and may be associated with a swimming pool, showers or bathing facilities. Such premises should be considered as a whole, and it must be borne in mind that the sauna cabin may well be located within the zones of the swimming pool.

The requirements for basic protection and fault protection in saunas are similar to those for bathrooms and swimming pools.

3.5 Wiring system

703.52 The wiring system should preferably be installed outside the zones, i.e. on the cold side of the thermal insulation. If the wiring system is installed on the warm side of the thermal insulation in zones 1 or 3, it must be heat-resisting. Any metallic sheaths or metallic conduits must not be accessible to persons using the sauna.

3.6 Heating elements

703.55 Sauna heating appliances must comply with BS EN 60335-2-53 and be installed in accordance with the manufacturer's instructions.

The heating elements incorporated in a sauna are likely to be metal sheathed. These, unless specified as having waterproof seals, may absorb moisture and cause the operation of a 30 mA RCD, if installed. RCD protection need not be provided for the sauna heater unless such protection is recommended by the manufacturer (see 3.4).

Construction and demolition site installations

4

4.1 Introduction

Sect 704 Section 704 of the 18th Edition includes minor changes to the requirements for external influences, and an addition to Regulation 704.410.3.6 concerning the protective measure of electrical separation.

4.2 Scope

704.1 Section 704 applies to all sites of construction work, including the repair or alteration of existing buildings and demolition work.

The requirements apply to fixed and movable installations.

The regulations do not apply to installations in administrative locations of construction sites (e.g. offices, cloakrooms, meeting rooms, canteens, restaurants, dormitories, toilets), where the general requirements of Parts 1 to 6 of BS 7671 apply.

4.3 The risks

Construction sites are potentially dangerous in many ways, but only those dangers that are associated with the risks of electric shock and burns are considered here. The risk of electric shock is high on a construction site because:

(a) of the possibility of damage to cables and equipment;
(b) of the wide use of hand tools with trailing leads;
(c) of the accessibility of many extraneous-conductive-parts, which cannot practically be bonded; and
(d) the works are generally open to the elements.

704.410.3.5 Section 704 prohibits the protective measures of obstacles and placing out of reach
704.410.3.6 (Section 417), non-conducting location (Regulation 418.1), and earth-free local equipotential bonding (Regulation 418.2).

In addition, Section 704 now makes it clear that electrical separation for the supply of more than one item of current-using equipment (Regulation 418.3) shall not be used.

4.4 Supplies

704.411.3.1 Regulation 704.411.3.1 states that a PME earthing facility shall not be used for the means of earthing for an installation falling within the scope of this section unless all extraneous-conductive-parts are reliably connected to the main earthing terminal in accordance with Regulation 411.3.1.2. A note in the regulation also refers the reader to BS 7375 (*Distribution of electricity on construction and demolition sites. Code of practice.*).

It is usually impracticable with PME to comply with the bonding requirements of the Electricity Safety, Quality and Continuity Regulations 2002 (ESQCR) on construction sites, and a PME earthing terminal should not be provided.

Where a construction site is part of an existing building and the building is supplied using a PME earthing facility the construction site will have to be separated from the PME building earth and be part of a TT system having a separate connection to earth that is independent from the PME building earth, unless the bonding requirements of Regulation 704.411.3.1 can be met. Where the construction site earthing and the building earthing are separate the earthing arrangements must not be simultaneously accessible.

The work should be planned in accordance with the Construction (Design and Management) Regulations 2015.

A note has been added to Regulation 704.313.3 pointing out that a single construction site may be served by several sources of supply and gives an example of generating sets.

4.5 Protection against electric shock

Section 704 prohibits the protective measures of obstacles and placing out of reach (Section 417), non-conducting location (Regulation 418.1) and earth-free local equipotential bonding (Regulation 418.2). In addition, Section 704 now makes it clear that the protective measure of electrical separation for the supply of more than one item of current-using equipment (Regulation 418.3) shall not be used.

4.5.1 Supplies to socket-outlets and hand-held equipment

Supplies to socket-outlets, etc. must comply with the requirements of Regulation 704.410.3.10:

704.410.3.10 A circuit supplying a socket-outlet with a rated current up to and including 32 A and any other circuit supplying hand-held electrical equipment with rated current up to and including 32 A shall be protected by:

(i) reduced low voltage (Regulation 411.8); or

(ii) automatic disconnection of supply (Section 411) with additional protection provided by an RCD having the characteristics specified in Regulation 415.1.1; or

(iii) electrical separation of circuits (Section 413), each socket-outlet and item of hand-held electrical equipment being supplied by an individual transformer or by a separate winding of a transformer; or

(iv) SELV or PELV (Section 414).

Where electrical separation is used, special attention should be paid to the requirements of Regulation 413.3.3.

BS 7671 strongly prefers the reduced low voltage system to supply portable handlamps for general use and portable hand tools and local lighting up to 2 kW, while SELV is strongly preferred for portable handlamps in confined or damp locations.

4.6 Reduced low voltage

411.8 110 V reduced low voltage supplies, with the centre point of the secondary winding of the step-down transformer earthed, limit the voltage to Earth to 55 V for single-phase supplies and 63.5 V for three-phase.

Limiting the voltage to 55 V or 63.5 V between a live conductor and Earth effectively eliminates the risk of dangerous electric shock from exposed-conductive-parts. Figure 4.1 shows single-phase and three-phase reduced low voltage supplies.

▼ **Figure 4.1** Reduced low voltage supplies

(a)

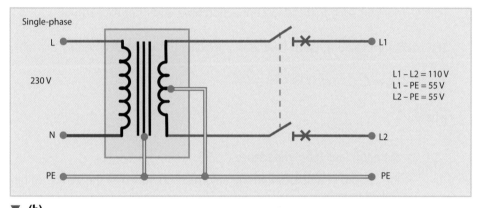

▼ **(b)**

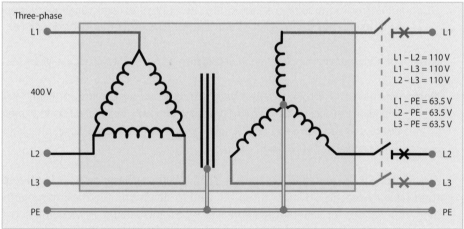

704.511.1 The movable reduced low voltage 110 V installation equipment is required to comply with BS EN 61439-4. 110 V plugs and sockets to BS EN 60309-2 (or BS 4343) are coloured yellow as recommended by the code of practice for distribution of electricity on construction and building sites (BS 7375). It is usual practice for the cables in the movable installation to have yellow sheaths to identify the voltage being used, although this is not a requirement of BS 7671 or BS 7375. Cable manufacturers produce arctic grade yellow PVC sheathed flexible cables to Table 6 of BS 6004:2012 for use in low temperature situations to distinguish from normal PVC sheathed cables.

It should be noted that whilst BS 7671 strongly prefers reduced low voltage systems for portable handlamps for general use and portable hand tools and local lighting up to 2 kW, the Health and Safety Executive have advised that the use of 230 V hand tools individually protected by 30 mA RCD may be acceptable.

GN5
GN8 The notes of guidance to the ESQCR advise that special consideration be given to the earthing and protection arrangements. Guidance Note 5: *Protection Against Electric Shock* and Guidance Note 8: *Earthing & Bonding* include guidance on earthing.

In addition to reduced low voltage, BS 7375 also provides information on typical distribution voltages for particular applications, for example, 400 V three-phase for fixed or movable plant over 3.75 kW and 230 V single-phase for fixed floodlighting and site building installations.

4.6.1 Maximum earth fault loop impedances

411.8.3 The value of earth fault loop impedance at every point of utilization, including socket-outlets, must be such that the disconnection time does not exceed 5 s.

The maximum value of earth fault loop impedance Z_s is determined by the formula:

$$Z_s \times I_a \leq U_0 \times C_{min} \quad \text{i.e} \quad Z_s \leq \frac{U_0 \times C_{min}}{I_a}$$

where:

Z_s is the earth fault loop impedance.

I_a is the current in amperes causing automatic operation of the device within the specified time.

U_0 is the nominal line voltage to Earth in volts.

C_{min} is the minimum voltage factor to take account of voltage variations depending on time and place, changing of transformer taps and other considerations.

Note: For a low voltage supply given in accordance with the ESQCR, C_{min} is given the value 0.95.

Note: Appendix 3 of BS 7671:2018 advises that the maximum values of earth fault loop impedance to achieve the disconnection time vary with the different types of protective device and also between manufacturers. Wherever possible, designers should use the manufacturer's specific data.

Alternatively, if the nominal voltage U_0 is 55 V single-phase or 63.5 V three-phase then Table 41.6 of BS 7671:2018 can be used to determine the maximum earth fault loop impedances for various circuit-breaker types and ratings. However, it should be noted that the values given are not to be exceeded when the line conductors are at the appropriate maximum permitted operating temperature and the circuit protective conductors are at the appropriate assumed initial temperature.

If general-purpose fuses to BS HD 60269-2:2013 or BS 88-2:2013 are used and the nominal voltage U_0 is 55 V single-phase or 63.5 V three-phase, then Table 41.6 of BS 7671:2018 can again be used to determine the applicable maximum earth fault loop impedances.

For a fuse of a different type and rating, the appropriate British Standard should be referenced to determine the value of I_a for the disconnection time in accordance with the appropriate value of the nominal voltage U_0. This can then be used to determine the maximum earth loop impedance to comply with $Z_s \times I_a \leq U_0 \times C_{min}$.

4.7 External influences

704.512.2 New Regulation 704.512.2 requires consideration to be given to the risk of damage to electrical equipment by corrosive substances, movement of structures and vehicles, wear and tear, tension, flexing, impact, abrasion, severing and ingress of liquids or solids.

4.8 Wiring systems

704.52 Cables on a construction site location should preferably not be installed across walkways or site roads as they are susceptible to mechanical damage. If cables are installed in this manner they would require the appropriate level of protection against mechanical damage and contact with construction plant machinery. Surface-run and overhead cables must be protected against mechanical damage, taking into account the environment and activities of a construction site.

For reduced low voltage systems, low temperature 3182/3/4/5A thermoplastic cable (to Table 6 of BS 6004) or equivalent flexible cable should be used. These cables remain flexible at lower temperatures than standard PVC cables and are ideal for outdoor use. They are referred to as arctic grade cable and typically have yellow (refer to section 4.6) or blue sheaths.

For cables used for applications exceeding reduced low voltage, flexible cables rated at 450/750 V that are resistant to abrasion and water should be used, type H07RN-F (BS EN 50525-2-21) or equivalent. These are heavy duty rubber insulated and sheathed flexible cables suitable for outdoor use.

4.9 Isolation and switching

704.537.2 Regulation 704.537.2 requires each Assembly for Construction Sites (ACS) to incorporate suitable devices for the switching and isolation of the incoming supply.

A device for isolating the incoming supply must be suitable for securing in the off position, e.g. by padlock or location of the device inside a lockable enclosure.

Current-using equipment must be supplied by ACSs (see Figure 4.2), each ACS comprising:

(a) overcurrent protective devices;
(b) devices affording fault protection; and
(c) socket-outlets, if required.

Safety and standby supplies must be connected by means of devices arranged to prevent interconnection of the different supplies.

▼ **Figure 4.2** Examples of assemblies for construction sites

▼ **(a)** Typical power assembly   **(b)** Typical lighting assembly

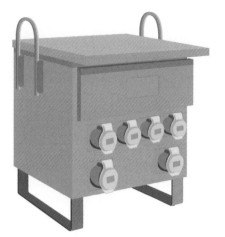 

4.10   Protection against the weather and dust

Appx 5 BS 7671   All equipment that is part of the movable installation should have a degree of protection appropriate to the external influences. Equipment for external use should be at least IP44. However, equipment installed in a weather protected location, such as an office being refurbished, would have no specific IP requirement.

4.11   Inspection and testing

GN3   It is recommended that the maximum period between inspections of construction site installations is 3 months.

Fixed installation RCDs should be tested daily (using the integral test button). Should RCDs be used to protect mobile equipment they must be tested by the operative before each period of use (again using the integral test button) and by the responsible person every 3 months (using an RCD tester).

Recommended intervals for inspection and testing of equipment are given in the IET *Code of Practice for In-service Inspection and Testing of Electrical Equipment.*

Agricultural and horticultural premises

5

5.1   Introduction

Sect 705   There are no significant changes to Section 705 introduced by the 18th Edition.

5.2   Scope

The particular requirements of Section 705 apply to fixed electrical installations indoors and outdoors in agricultural and horticultural premises. Some of the requirements are also applicable to other locations that are in common buildings belonging to the agricultural and horticultural premises. Where special requirements also apply to residences and other locations in such common buildings this is stated in the text of the relevant regulations.

Rooms, locations and areas for household applications and similar are not covered by this section.

Note: Chapter 5 does not cover electric fence installations, for which reference should be made to BS EN 60335-2-76.

5.2.1   Residences and other locations belonging to agricultural and horticultural premises

Part 2   These are defined as:

Residences and other locations which have a conductive connection to the agricultural and horticultural premises by either protective conductors of the same installation or by extraneous-conductive-parts.

Note: Examples of other locations include offices, social rooms, machine-halls, workrooms, garages and shops.

5.3   The risks

The particular risks associated with farms and horticultural premises are:

(a) general accessibility of extraneous-conductive-parts;
(b) an onerous environment with respect to mechanical damage, exposure to the weather, corrosive effects (from water, animal urine, farm chemicals etc.);
(c) a mechanically hazardous area due to electromechanical equipment, mills and mixers, and mechanical drives of all kinds;
(d) rodent damage to (gnawing of) cables, leading to fire risks;
(e) storage of flammable materials, e.g. straw and grain; and
(f) increased susceptibility of electric shock for livestock.

5.4 Electricity supplies

Because of the practical difficulties in bonding all accessible extraneous-conductive-parts, electricity distribution companies might not provide a PME earth to agricultural or horticultural installations.

The Department for Business, Energy and Industrial Strategy (formerly the DTI), guidance on the ESQCR {9(4)} advises that special consideration should be given to the earthing and bonding requirements for farms, where it may prove difficult to attach and maintain all the necessary equipotential bonding connections for a PME supply.

705.415.2.1 Regulation 705.415.2.1 refers to Figure 705, which shows the requirements for bonding in cattle sheds. PME is **not** recommended if there is no metal grid laid in the floor.

A TN-S supply is unlikely to be provided by a distributor as a routine unless the installation is particularly large and early application is made. Alternatively, consideration should be given to installing an additional earth electrode, as it is most likely that the installation will be required to be TT.

It should be noted that in TT installations isolators are required to switch all live conductors, including the neutral.

705.411.4 Regulation 705.411.4 prohibits the use of a TN-C system.

5.5 Protection against electric shock

705.410.3.5 The protective measures of obstacles and placing out of reach (Section 417) shall not be used.

705.410.3.6 The protective measures of non-conducting location (Regulation 418.1) and earth-free local equipotential bonding (Regulation 418.2) shall not be used.

5.5.1 Protective measure: Automatic disconnection of supply

705.411.1 RCDs must be provided for automatic disconnection of supply to circuits as follows (see Figure 5.1):

705.411.1 In circuits, whatever the type of earthing system, the following disconnection devices shall be provided:

415.1.1 **(i)** in final circuits supplying socket-outlets with rated current not exceeding 32 A, an RCD having the characteristics specified in Regulation 415.1.1;
(ii) in final circuits supplying socket-outlets with rated current more than 32 A, an RCD with a rated residual operating current not exceeding 100 mA; and
(iii) in all other circuits, RCDs with a rated residual operating current not exceeding 300 mA.

▼ **Figure 5.1** Requirements for RCDs

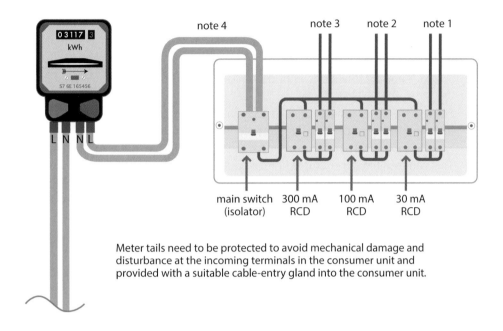

main switch 300 mA 100 mA 30 mA
(isolator) RCD   RCD   RCD

Meter tails need to be protected to avoid mechanical damage and disturbance at the incoming terminals in the consumer unit and provided with a suitable cable-entry gland into the consumer unit.

Notes:

(i) Socket-outlet circuits not exceeding 32 A.

(ii) Socket-outlet circuits exceeding 32 A.

(iii) Circuits other than socket-outlets.

531.3.5.3.1 **(iv)** The risk of faults to exposed-conductive-parts on the supply side of the main switch must be minimized because such faults would not be detected by the RCDs. For RCDs in a TT system see Regulation 531.3.5.3.1. Where a metalclad consumer unit is installed on a TT system the main double-pole isolator on the consumer unit may need to be a time-delayed RCD, depending on the construction of the consumer unit. Meter tails should meet the requirements of double or reinforced insulation. Meter tails need to be protected against mechanical damage to avoid the risk of disturbance of the incoming connections to the consumer unit and therefore prevent a conductor becoming disconnected and making contact with the metal case of the consumer unit. Particular attention should be placed on the entry point of the meter tails into a metal consumer unit to avoid damage to the meter tails.

PD 6519-3:1999 (IEC A BSI published document PD 6519-3 (IEC 60479-3) *Guide to effects of current on*
60479-3:1998) *human beings and livestock – Part 3 Effects of current passing through the body of livestock* is available that provides guidance on the effects of electric currents on livestock. The report indicates values for the impedance of the body of livestock as a function of the touch voltage, the degree of moisture of the hide or skin and the current path.

The report includes the following information:

▶ the internal body impedance is considered as mostly resistive and depends primarily on the current path.

▶ the impedance of a hide depends largely on humidity. In dry conditions the hide can be considered to be practically an insulator for voltages up to 100 V with impedance values in the range of tens to hundreds of kilohms.

- the impedance of skin depends on the voltage, frequency, duration of current flow, surface area of contact, pressure of contact, degree of moisture of the skin and temperature. The skin impedance falls as the current is increased.
- the impedance of the hind legs is smaller than the impedance of the forelegs. The impedance from the nose to all four legs is smaller than the impedance of forelegs to hind legs.
- the total body impedance (cattle) for a current path nose to the four legs is typically 35 % of the total body impedance for a current path forelegs to hind legs.

However, animals will detect relatively small voltage gradients between front and rear legs, and between conductive part potentials and Earth.

705.415.2.1 This can result in, for example, a marked reluctance for cows to enter milking parlours because of potential differences. These potential differences can arise from a number of causes. If the electricity supply is PME, at the end of a long run there is likely to be a potential between the PME earth and true Earth. In locations intended for livestock, supplementary protective equipotential bonding must connect all exposed-conductive-parts and extraneous-conductive-parts that can be touched by livestock. Where a metal grid is laid in the floor, it must be included within the supplementary bonding of the location (Figure 705 of BS 7671 shows an example of this; other suitable arrangements of a metal grid are not precluded). Extraneous-conductive-parts in, or on, the floor, e.g. concrete reinforcement in general or reinforcement of cellars for liquid manure, must be connected to the supplementary protective equipotential bonding. It is recommended that spaced floors made of prefabricated concrete elements be part of the supplementary equipotential bonding. The supplementary bonding and the metal grid, if any, must be erected so that it is durably protected against mechanical stresses and corrosion.

Note: Where a metal grid is not laid in the floor a PME supply is not recommended. It is well known that animals have received shocks that are associated with electrical installations. It is likely that lightning strikes on overhead lines conducted to Earth via earth electrodes at the bottom of a pole produce voltage gradients that are fatal to animals because of the wide spacing of their feet. If there is concern in this respect, the location of earth electrodes should be discussed with the electricity distribution company.

5.6 Earth fault loop impedances

411.5.3 Where an RCD is used for earth fault protection, the following conditions are to be fulfilled for a TT installation:

The disconnection time must satisfy Regulation 411.3.2.2 or 411.3.2.4, and $R_A \times I_{\Delta n} \leq 50$ V

where:

R_A is the sum of the resistances of the earth electrode and the protective conductor connecting it to the exposed-conductive-parts (in ohms).

$I_{\Delta n}$ is the rated residual operating current of the RCD.

The above conditions would be met where the earth fault loop impedance does not exceed the values stated in Table 5.1.

▼ **Table 5.1** Maximum earth fault loop impedance

Rated residual operating current, $I_{\Delta n}$ (mA)	Maximum earth fault loop impedance, Z_S (ohms)
30	1667*
100	500*
300	167

* The resistance of the installation earth electrode itself should be as low as practicable. A value exceeding 200 ohms may not be stable. Refer to Regulation 542.2.4.

5.7 Protection against fire

Regulation 705.422.6 requires electrical heating appliances used for the breeding and rearing of livestock to comply with BS EN 60335-2-71 and to be fixed so as to maintain an appropriate distance from livestock and combustible material, to minimize any risks of burns to livestock and of fire. For radiant heaters the clearance should be not less than 0.5 m or such other clearance as recommended by the manufacturer. For luminaires refer to Regulations 422.3.1 and 422.4.2.

705.422.7 Regulation 705.422.7 requires that for additional fire protection purposes in some circumstances, RCDs are to be installed with a rated residual operating current not exceeding 300 mA and must disconnect all live conductors. Where improved continuity of service is required, RCDs not protecting socket-outlets need to be of the S type or have a time delay (see Figure 5.2).

705.422.8 Regulation 705.422.8 specifies requirements for conductors of circuits supplied at extra-low voltage in locations where a fire risk exists.

Rodent damage is a major cause of farm fires and this must be taken into account by the designer and installer. Cables should be installed and routed with such potential damage in mind. For example, in a livestock building, cables should be routed on the underside of the ceiling rather than in a false roof. Steel conduit provides a good degree of protection.

Additional protection to mitigate against the risk of fire due to arcing is recommended.

421.1.7 Regulation 421.1.7 recommends the use of arc fault detection devices (AFDDs) conforming to BS EN 62606 as a means of providing additional protection against fire caused by arc faults in AC final circuits.

If used, an AFDD should be placed at the origin of the circuit to be protected.

Notes:
Examples of where such devices can be used include:
(a) locations with a risk of fire due to the nature of processed or stored materials, i.e. BE2 locations (e.g. barns, woodworking shops, stores of combustible materials); and
(b) locations with combustible constructional materials, i.e. CA2 locations (e.g. wooden buildings).

AFDDs conforming to BS EN 62606 protect against series and parallel arcing and are designed to detect low-level hazardous arcing that circuit-breakers, fuses and RCDs are not designed to detect. Such arcing could be caused by factors such as rodents gnawing cables (leading to fire risks) or storage of flammable materials, e.g. straw and grain.

GN4   Additional guidance for protection against fire for this type of location can be found in IET Guidance Note 4: *Protection Against Fire*.

▶ **Figure 5.2** Example schematic of supply arrangements (including livestock support)

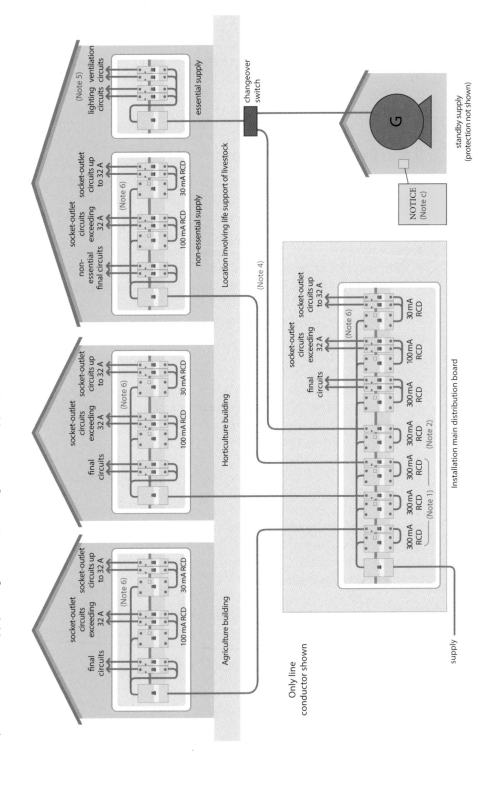

Notes to Figure 5.2:

(a) All circuits other than socket-outlets must be protected by 300 mA RCDs for fire protection purposes. These will need to discriminate with final circuit RCDs where relevant.

(b) For improved continuity of service, RCDs should be time-delay or S type as long as they are not supplying socket-outlets.

(c) A standby supply should have a notice adjacent to it to highlight that it should be periodically tested in accordance with the manufacturer's instructions.

(d) Discrimination of main circuits supplying ventilation for life support circuits.

(e) Separation of lighting and ventilation circuits for life support.

(f) Circuits supplying socket-outlets up to 32 A require protection by 30 mA RCDs. Circuits supplying socket-outlets exceeding 32 A require protection by 100 mA RCDs.

5.8 External influences

705.512.2 Electrical equipment is required to have a minimum degree of protection of IP44 rating, when used under normal conditions. Where equipment of IP44 rating is not available, it should be placed in an enclosure complying with IP44.

- ▶ IPX4 provides protection against water splashing;
- ▶ IPX5 provides protection against water jets from any direction; and
- ▶ IPX6 provides protection against powerful water jets from any direction.

GN1 See Guidance Note 1: *Selection & Erection* for details of degrees of protection provided by enclosures.

Socket-outlets should be installed in a position where they are unlikely to come into contact with combustible material. Where there are conditions of external influences >AD4, >AE3 and/or >AG1, socket-outlets must be provided with the appropriate protection. Protection may also be provided by the use of additional enclosures or by installation in building recesses.

See Appendix 5 of BS 7671 for classification of external influences.

The foregoing requirements do not apply to residential locations, offices, shops and locations with similar external influences belonging to agricultural and horticultural premises where, for socket-outlets, BS 1363-2 or BS 546 applies. Where corrosive substances are present, e.g. in dairies or cattle sheds, the electrical equipment needs to be adequately protected.

Luminaires should comply with the BS EN 60598 series and be selected for their degree of protection against the ingress of dust, solid objects and moisture (e.g. IP54).

Table 55.3 Luminaires marked ▽D̷ have a limited surface temperature of the luminaire, and should have a degree of protection of IP54.

5.9 Wiring systems

705.522 Regulation 705.522 calls for wiring systems to be inaccessible to livestock or suitably protected against mechanical damage.

Where vehicles and mobile agricultural machines are operated, underground cables must be buried in the ground at a depth of at least 0.6 m with added mechanical protection. Cables in arable or cultivated ground should be buried at a depth of at least 1 m.

Self-supporting suspension cables should be installed at a height of at least 6 m.

705.522.16 Regulation 705.522.16 has requirements for conduit, cable trunking and ducting systems to be protected against corrosion and impact.

5.10 Safety services

705.560.6 Where an installation includes high density livestock rearing, there may be a need to take account of the continued operation of systems for the livestock (see Figure 5.2).

If the supply of food, water, air and/or lighting to the livestock is not provided in the event of a power supply failure then a separate source of supply should be provided. This would include either an alternative supply or back-up supply.

For the supply of ventilation and lighting units, separate final circuits should be provided. These circuits should be limited to supporting just the essential equipment of the ventilation and lighting. Selectivity of the circuits supplying ventilation should be provided.

Electrically powered ventilation should include either of the following:

▶ a standby electrical source of supply that can ensure the operation of the ventilation equipment. If a standby source is used then a notice adjacent to this source should be provided highlighting that the standby source should be tested periodically in accordance with the manufacturer's instructions.
▶ monitoring of the temperature and supply voltage by one or more devices. These should provide a visual or audible signal that is located in a position that can be readily observed by the user, and the device(s) should operate independently from the normal supply.

5

Guidance Note 7: Special Locations
© The Institution of Engineering and Technology

Conducting locations with restricted movement   6

6.1   Introduction

Sect 706   There are no significant changes to Section 706 introduced by the 18th Edition.

6.2   Scope

Part 2
706.1   A conducting location with restricted movement is constructed mainly of metallic or other conductive surrounding parts, and within it movement is restricted. It is likely that a person in such a location will be in good contact with conductive surroundings and escape will be difficult in the event of an electric shock. The regulations apply to the fixed conducting location and to the supplies for mobile and hand-held equipment for use in such locations.

The particular requirements of Section 706 do not apply to locations that allow persons freedom of bodily movement, that is, to enter, work and leave the location without physical constraint. The types of location that are being considered include boiler shells, cable gantries, small tunnels, metal sewers, etc.

For installation and use of arc-welding equipment, see IEC 60974-9.

This section does not apply to electrical systems as defined in BS 7909 used in structures, sets, mobile units etc. as used for public or private events, touring shows, theatrical, radio, TV or film productions and similar activities of the entertainment industry.

6.3   The risks

In many ordinary locations, there is usually limited access to earthed metal. As a result, the likelihood of receiving a shock current of sufficient magnitude to have harmful physiological effects is low. This is not so with a conducting location.

In a conducting location where bodily movement is limited, there is little opportunity to move away from the shock. Contact resistance is low due to high contact areas and perspiration, so that body currents are high and the risk of ventricular fibrillation is also high. There are other effects of electric shock that are also relevant in such locations. Muscles used to breathe can be constrained by body currents, and shocks to the head can paralyze the breathing function.

6.4 Protection against electric shock

706.410.3.5 The protective measures of obstacles and placing out of reach (Section 417) shall not be used in these locations.

706.410.3.10 For supplies to equipment, the following protective measures are permissible:

Hand-held tools or items of mobile equipment

(a) electrical separation, only one item of equipment being connected to a secondary winding of the transformer (which may have two or more such windings); or

(b) SELV.

Handlamps

(a) SELV.

Fixed equipment

(a) automatic disconnection of supply with supplementary equipotential bonding connecting exposed-conductive-parts of fixed equipment and the conductive parts of the location; or

415.1.1 **(b)** use of Class II equipment or equipment having equivalent insulation, provided the supply circuits have additional protection because they have RCDs having the characteristics specified in Regulation 415.1.1; or

(c) electrical separation, only one item of equipment being connected to a secondary winding of the transformer; or

(d) SELV or PELV.

For PELV, protective equipotential bonding must be provided between all exposed-conductive-parts and extraneous-conductive-parts inside the location and the connection of the PELV system to Earth.

706.413 For electrical separation, in addition to the requirements of Section 413, the unearthed source should have simple separation and should be situated outside the conducting location with restricted movement, unless the source is part of the fixed electrical installation of the location.

706.414 For SELV or PELV, the general requirements of Section 414 apply, except that, whatever the nominal voltage, basic protection must be provided. In addition, the source of SELV or PELV must be situated outside the conducting location with restricted movement, unless it is part of the fixed installation.

Electrical installations in caravan/camping parks, caravans and motor caravans

7

7.1 Introduction

Sect 708   The scope of Section 708 has been extended to cover circuits intended to supply residential park homes in caravan parks, camping parks and similar locations. In addition, changes have been made to socket-outlet requirements, RCD protection and external influences.

Sect 721   Section 721 contains a number of changes, including requirements for electrical separation, RCDs, proximity to non-electrical services and protective bonding conductors.

Arrangement of the 18th Edition

Sect 708   The 18th Edition has two separately numbered sections, namely Section 708
Sect 721   concerning the electrical installations in caravan parks, etc., and Section 721 concerning caravans and motor caravans.

This chapter provides guidance on applying the particular requirements of both Sections 708 and 721 of BS 7671:2018.

7.2 Scope of Sections 708 and 721

7.2.1 Section 708

708.1   The particular requirements of this section apply to that portion of the electrical installation in caravan/camping parks and similar locations providing facilities for supplying leisure accommodation vehicles (including caravans), tents or residential park homes.

The requirements do not apply to the internal electrical installations of leisure accommodation vehicles, mobile or transportable units, or residential park homes.

7.2.2   Section 721

721.1   The particular requirements of this section apply to the electrical installations of caravans and motor caravans at nominal voltages not exceeding 230/400 V AC or 48 V DC, except for 12 V DC – see exclusions below.

The requirements do apply to electrical circuits and equipment intended for the use of the caravan for habitation purposes.

The requirements do not apply to the following:

(a) electrical circuits and equipment for automotive purposes;
(b) 12 V DC installations, which are covered by BS EN 1648-1 and BS EN 1648-2;
(c) electrical installations of mobile homes or residential park homes, to which the general requirements apply; and
(d) transportable units (see Chapter 14 of this Guidance Note).

For the purposes of Section 721, caravans and motor caravans are referred to as 'caravans'. The particular requirements of some other sections of Part 7 may also apply to electrical installations in caravans, e.g. Section 701 (showers).

7.3   The risks

The risks specifically associated with installations in caravan parks, caravans and motor caravans arise from:

(a) open-circuit faults of the PEN conductor of PME supplies that raise the potential to true Earth of all metalwork (including that of caravans if connected) to dangerous levels;
(b) incorrect polarity at the pitch supply point;
(c) inability to establish an equipotential zone external to the vehicle;
(d) possible loss of earthing due to long supply cable runs, connecting devices exposed to weather and flexible cable connections liable to mechanical damage; and
721.522.7.1 **(e)** vibration while the vehicle is moving, causing faults within the caravan installation.

Particular requirements to reduce the above risks include:

708.411.4   **(a)** prohibition of the connection of exposed- and extraneous-conductive-parts of a
708.553.1.14   caravan or motor caravan to a PME terminal. The protective conductor of each
411.5   socket-outlet to be connected to an earth electrode and the requirements of Regulation 411.5 for a TT system to be complied with.
721.415.1   **(b)** additional protection by 30 mA RCDs in both the vehicle and the park
708.415.1   installation.
721.537.2.1.1   **(c)** double-pole isolating switch and final circuit circuit-breakers in the caravan or
721.43.1   motor caravan.
721.524.1   **(d)** internal wiring of the caravan or motor caravan by flexible or stranded cables of
721.522.8.1.3   cross-sectional area 1.5 mm² or greater; additional cable supports; segregation
721.528.1   of low voltage and extra-low voltage circuits.

7.4   Legislation and standards

Regulation 9(4) of the ESQCR 2002 prohibits the connection of the supply neutral of a PME supply to any metalwork in a caravan (or boat).

Caravan parks (sites) in the United Kingdom are subject to the provisions of the Caravan Sites and Control of Development Act 1960. This empowers local authorities to issue licences and to impose conditions, generally in accordance with model standards. For residential parks there are the *Model Standards 2008 for Caravan Sites in England* and *Model Standards 2008 for Caravan Sites in Wales*. For holiday sites there are the *Model Standards 1989: Holiday Caravan Sites*. For touring sites there are the *Model Standards for Touring Caravan Sites*. These model standards include requirements for the electrical installation in the caravan park to be installed and maintained to the requirements of BS 7671.

The 1960 Act also empowers certain 'exempted bodies' such as the Caravan Club and the Camping and Caravanning Club to issue certificates in respect of parks for use by their own members.

While there is no legislation specific to electrical installations in caravans, BS EN 1645-1 *Leisure accommodation vehicles – caravans* includes a section on electrical installations. This requires that low voltage electrical installations are to comply with international standard IEC 60364-7-708 *Low voltage electrical installations – Part 7-708: Requirements for special installations or locations – Caravan parks, camping parks and similar locations*. These requirements are included in Section 708 of BS 7671.

Extra-low voltage installations using 12 V DC should comply with BS EN 1648 *Leisure accommodation vehicles – 12 V direct current extra-low voltage electrical installations – 1: Caravans*, or *– 2: Motor caravans*.

7.5   Electrical installations in caravan parks (708)

7.5.1   Supply systems

708.411.4 The supply system to the permanent buildings of a caravan park can be PME, TN-S or TT. Consideration should be given to the earthing and bonding of amenity buildings such as toilet and shower blocks. The supply system to the caravans is limited to TN-S or TT. The ESQCR prohibit the connection of a PME earthing facility to any metalwork in a leisure accommodation vehicle (caravan). If the caravan supply is derived from a permanent building that is supplied by a PME system then the caravan supply will have to be part of a TT system having a separate connection to Earth independent from the PME earthing.

The separation of the earthing should preferably be effected at the main distribution board (see Figure 7.1). This enables the exposed-conductive-parts connected to each system to be more readily identified and inspected periodically. An earth electrode for the TT system should be provided nearby and located so that the resistance areas of the PME supply earthing and earth electrode do not overlap (refer to section 7.5.6).

Alternatively, the separation of the earthing can be made at the caravan pitch supply points. In this instance, earth electrodes will be required at these points. A cable supplying a separate earthing system is to be earthed only at the installation containing the associated protective device (see Figure 7.2).

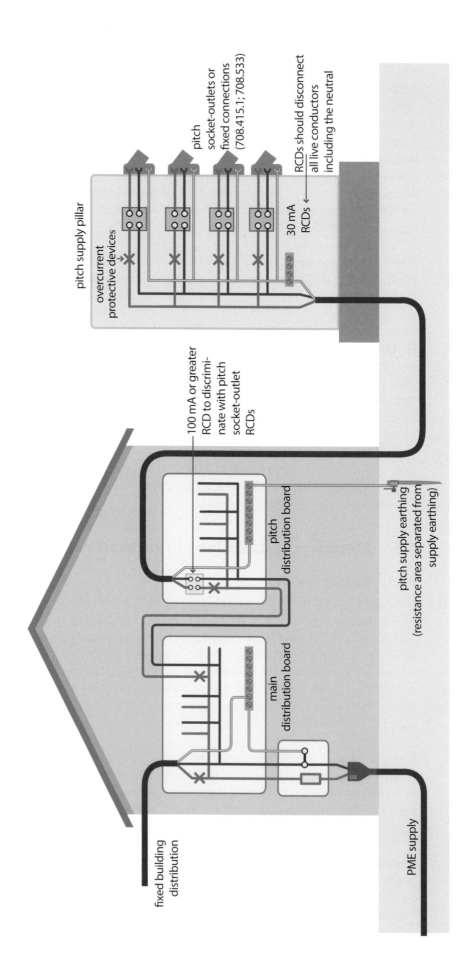

▶ **Figure 7.1** Typical site distribution for a PME supply, separation from PME earth at main distribution board

▶ **Figure 7.2** Typical site distribution for a PME supply, separation from PME earth at pitch supply point

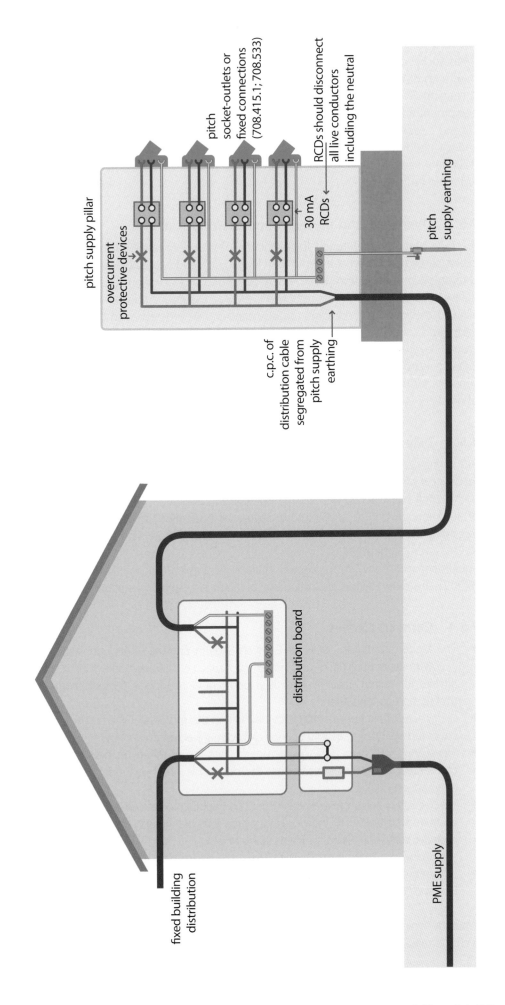

pitch
socket-outlets or
fixed connections
(708.415.1; 708.533)

RCDs should disconnect
all live conductors
including the neutral

pitch
supply earthing

pitch supply pillar

overcurrent
protective devices

30 mA
RCDs

c.p.c. of
distribution cable
segregated from
pitch supply
earthing

distribution board

fixed building
distribution

PME supply

411.3.2.2
Table 41.1
411.3.2.3
411.3.2.4
314.1(i)

For automatic disconnection of supply, final circuits not exceeding 63 A with one or more socket-outlets and 32 A supplying only fixed current-using equipment require maximum disconnection times of 0.4 s for TN and 0.2 s for TT systems. Final circuits exceeding 63 A serving socket-outlets and 32 A supplying only fixed current-using equipment and distribution circuits require maximum disconnection times of 5 s for TN and 1 s for TT systems. For TT supplies an RCD of 100 mA or more will be required and will need to be a delay or 'S' type to provide selectivity with the individual 30 mA RCDs required for the pitch outlets. If an RCD is used for a TN-S system then selectivity will again be required.

7.5.2   Overhead and underground distribution

708.521.7
708.410.3.5

BS 7671 states a preference for distribution by underground cables. Underground cables should be buried at a depth of at least 0.6 m and should be routed outside any caravan pitch or away from areas where tent poles or ground anchors could be used, unless additional mechanical protection is provided. Overhead distribution systems are allowed provided all conductors are insulated and so erected as to be unlikely to be damaged by vehicle movement. They are required to be not less than 6 m above ground level in areas subject to vehicle movement and 3.5 m in all other areas. Suitable warning notices should be displayed at the entrance to the site and on supports for the overhead line and, where appropriate, attention drawn to the danger of masts of yachts or dinghies contacting the overhead line.

7.5.3   External influences

708.512

Any wiring system or equipment selected and installed must be suitable and sufficient for its location and able to operate satisfactorily. Suitable protection must be provided, both during construction and for the completed installation. Regarding the presence of solid foreign bodies, a minimum degree of protection of IP4X is now required. Regarding the presence of water, a minimum degree of protection of IPX4 is required.

708.512.2.1.3

Equipment must now be protected against mechanical impact IK 08 (see BS EN 62262) and/or located to avoid damage by any reasonably foreseeable impact.

7.5.4   Caravan pitches

▶ The Model Standards 2008 for Caravan Sites in England – and similar for Wales – covers residential parks and mixed use parks, i.e. holiday parks with residential pitches for units/caravans used as a permanent residence protected under the mobile homes legislation.

▶ Dimensions of pitches are **NOT** provided within the model standards. The standards cover the boundaries, density, spacing and parking between caravans.

▶ The spacing/separation distances between units are required for health and safety considerations and for privacy from the other side of the boundary and between neighbouring units.

▶ The Regulatory Reform (Fire Safety) Order 2005 applies to caravan sites with common or shared parts and to individual caravans that are holiday-let, i.e. rented out. It overrides some fire-related standards that may be in current site licensing conditions, e.g. separation distances for units with lower fire-performance properties.

▶ Where the Order applies, the site owner must carry out a fire risk assessment. The Fire and Rescue Service can provide safety advice.

When designing pitches, the requirements of the above legislation should be confirmed with the relevant local authority.

7.5.5   Socket-outlets, overcurrent and RCD protection, and isolation

Socket-outlets

The requirements for socket-outlets have been enhanced to prevent the socket contacts being live when accessible.

708.55.1.1   Regulation 708.55.1.1 requires that every socket-outlet or connector shall either comply with:

(a) BS EN 60309-2 and shall be interlocked and classified to Clause 6.1.5 of BS EN 60309-1:1999, to prevent the socket contacts being live when accessible; or

(b) be part of an interlocked self-contained product complying with BS EN 60309-4 and classified to Clauses 6.1.101 and 6.1.102 of BS EN 60309-4:2006, to prevent the socket contacts being live when accessible.

708.55.1.3
708.55.1.4
708.55.1.5
708.553.1.8
The current rating is to be no less than 16 A, but may be greater if required. Caravan pitch socket-outlets are required to comply with BS EN 60309-2 and to have a degree of protection of at least IP44. At least one socket-outlet must be provided for each caravan pitch. Where socket-outlets are grouped in pitch supply equipment, there should be one socket-outlet for each pitch, limited to a group of four.

Overcurrent protection

Every socket-outlet shall be individually protected by an overcurrent protective device, in accordance with the requirements of Chapter 43 of BS 7671.

A fixed connection for a supply to a mobile home or residential park home shall be individually protected by an overcurrent protective device, in accordance with the requirements of Chapter 43.

RCD protection

Each socket-outlet must be protected individually by an RCD having the characteristics specified in Regulation 415.1.1 for additional protection. The RCD must disconnect all live conductors, including the neutral. (See Figures 7.1 and 7.2.)

Requirements for RCD protection have been extended to cover supplies to residential park homes. A final circuit (from the metering point) intended for the fixed connection for a supply to a mobile home or a residential park home shall be individually protected by an RCD having a rated residual operating current not exceeding 30 mA accessible to the consumer. Devices selected shall disconnect all live conductors. This requirement is to prevent a situation in which a park or mobile home resident is unable to re-energize a supply after a fault, owing to the RCD being incorrectly located in a locked, inaccessible or remote cabinet.

Isolation

708.537.2.1.1   Regulation 708.537.2.1.1 now requires at least one means of isolation to be installed in each distribution enclosure. This device shall disconnect all live conductors.

7.5.6   Separation of electrodes

Figure 7.3 indicates that effective separation of resistance areas of earth electrodes is achieved if the distance between the electrodes exceeds 10 m.

▼ **Figure 7.3** Ground surface potentials around a single rod and three rods in line

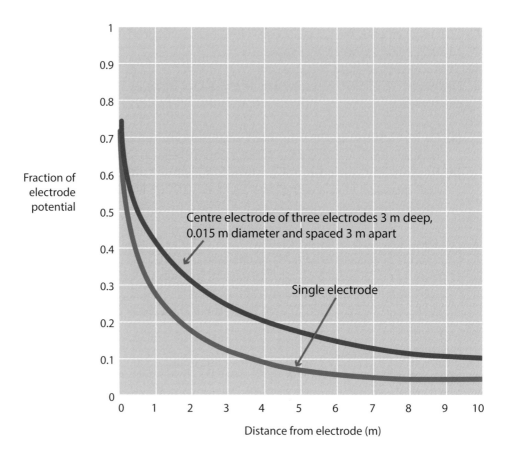

GN3 Guidance Note 3: *Inspection & Testing* describes a test method for the measurement of earth electrode resistance.

7.6   Caravans and motor caravans (721)

7.6.1   Protective equipotential bonding

721.411.3.1.2 Structural metallic parts that are accessible from within the caravan are required to be connected through main protective bonding conductors to the main earthing terminal within the caravan.

7.6.2   Provision of RCDs

721.415.1 The requirements for RCD protection have been redrafted. Regulation 721.415.1 states that where protection by automatic disconnection of supply is used, an RCD with a rated residual operating current not exceeding 30 mA, complying with BS EN 60947-2 (Annex B), BS EN 61008-1, BS EN 61009-1 or BS EN 62423 breaking all live conductors, shall be provided, having the characteristics specified in Regulation 415.1.1.

Each supply inlet shall be directly connected to its associated RCD. Please note that this implies that there may not be any taps or junctions in this connection.

The wiring system must include a circuit protective conductor connected to:

(a) the protective contact of the inlet;
(b) the exposed-conductive-parts of the electrical equipment; and
(c) the protective contacts of the socket-outlets.

7.6.3 Protective equipotential bonding

Regulation 721.411.3.1.2 Regulation 721.411.3.1.2 requires structural metallic parts that are accessible from within the caravan to be connected through main protective bonding conductors to the main earthing terminal within the caravan.

Regulation 721.544.1.1 The requirements for connections of protective bonding conductors have been clarified. Regulation 721.544.1.1 states that the terminations of protective bonding conductors connecting the conductive structure of the unit shall be accessible and protected against corrosion.

7.6.4 Protection against overcurrent

721.43.1 Each final circuit must be protected by an overcurrent protective device, which disconnects all the live conductors of the circuit.

7.6.5 Selection and erection of equipment

721.510.3 More than one electrically independent installation is permitted, provided that each
528.1 one is supplied via a separate connecting device and they are segregated in accordance with the requirements of Regulation 528.1. This includes requirements for insulating cables or conductors to the highest voltage present, or installing the independent circuit cables in separate conduits, trunking or ducting compartments.

This enables 12 V DC battery-supplied circuits for interior lighting in accordance with BS EN 1648 *Leisure accommodation vehicles 12 V direct current extra-low voltage installations, Part 1 – caravans, Part 2 – motor caravans* and for the road lighting circuits of the caravan or motor caravan in accordance with the correct amendment of the Road Vehicle Lighting Regulations 1989.

7.6.6 Proximity to non-electrical services

The requirements for proximity to non-electrical services have been redrafted.

Regulation 721.528.2.1 Regulation 721.528.2.1 requires that where cables must run through a gas cylinder storage compartment, they shall pass through the compartment at a height of not less than 500 mm above the base of the cylinders. They shall be protected against mechanical damage by installation within a conduit system complying with the appropriate part of the BS EN 61386 series or within a ducting system complying with the appropriate part of the BS EN 50085 series.

7.6.7 Switchgear and controlgear

721.537.2.1.1 The installation to the caravan should have a main disconnector, which will disconnect all the live conductors. This should be placed in a suitable position for ready operation within the caravan to isolate the supply. When a caravan only has one final circuit then the isolation can be afforded by the overcurrent protective device as long as it fulfils the requirements for isolation.

Figure 721 An indelible notice in the appropriate language(s) must be permanently fixed near the main isolation point inside the caravan to provide the user with instructions on connecting and disconnecting the supply (refer to Figure 721 of BS 7671).

721.55.1 The inlet to the caravan must be an appliance inlet complying with BS EN 60309-1. This should be installed not more than 1.8 m above ground level, in a readily accessible position, have a minimum degree of protection of IP44, and should not protrude significantly beyond the body of the caravan.

7.6.8 The connecting flexible cable

721.55.2.6 The means of connecting the caravan to the pitch socket-outlet should be provided with the caravan. This must have a plug at one end complying with BS EN 60309-2, a flexible cable with a continuous length of 25 m (±2 m). The connecting flexible cable must be in one length, without signs of damage, and not contain joints or other means to increase its length; and a connector if needed that is compatible with the appropriate appliance inlet. The cable should be to the harmonized code designation H05RN-F or H07RN-F (BS EN 50525-2-21) or equivalent, include a protective conductor, have cores coloured as required by Table 51 of BS 7671 and have a cross-sectional area as shown in Table 7.1.

Note: The National Caravan Council (NCC) and the wider caravan industry regard H05VV-F as being equivalent to H05VV-F.

▼ **Table 7.1** Minimum cross-sectional areas of flexible cables for caravan connection (Table 721 of BS 7671:2018)

Rated current (A)	Minimum cross-sectional area (mm²)
16	2.5
25	4
32	6
63	16
100	35

7.7 Tents

708.553.1 Mains supplies to tents (marquees, etc.) should generally incorporate the equipment
708.415.1 specified for a caravan, namely a main double-pole isolating switch and a main double-pole RCD (which may be combined), one or more double-pole circuit-breakers, and socket-outlets as required. Double-pole circuit-breaker means having overcurrent detection and switching in both poles and not a single-pole circuit-breaker with a switched neutral. The connecting cable should be securely clamped to the main supply unit and directly connected to it without the use of a plug and inlet. Luminaires and appliances should be of Class II construction.

If the tent or marquee is used for a temporary event, such as a wedding reception, refer to BS 7909 *Code of practice for temporary electrical systems for entertainment and related purposes.*

Marinas and similar locations   **8**

8.1 Introduction

Sect 709   For electrical installations of pleasure craft, reference should be made to BS EN 60092-507 *Electrical installations in ships – Pleasure craft*, while for houseboats the general requirements of BS 7671 apply.

There are no significant changes to Section 709 introduced by the 18th Edition.

8.2 Scope

709.1   The particular requirements of Section 709 are applicable only to circuits intended to supply pleasure craft or houseboats in marinas and similar locations. (In this section 'marina' means 'marina and similar locations'.)

The particular requirements do not apply to the supply to houseboats if they are supplied directly from the public network, or to the internal electrical installations of pleasure craft or houseboats.

For the remainder of the electrical installation of marinas and similar locations the general requirements of the Regulations apply, together with any relevant particular requirements of Part 7.

8.3 The risks

709.512.2   The environment of a marina or yachting harbour is harsh for electrical equipment. The water, salt and movement of structures accelerate deterioration of the installation. The presence of salt water, dissimilar metals and a potential for leakage currents increases the rate of corrosion. There are also increased electric shock risks associated with a wet environment because of reduction in body resistance and contact with earth potential.

Site investigations should be carried out at an early stage to determine likely maximum wave heights. This is of particular importance in exposed coastal sites. Where marinas have breakwater type pontoons, it is likely that under certain conditions waves will pass over the structure.

The risks specifically associated with craft supplied from marinas include:

(a) open-circuit faults of the PEN conductor of PME supplies raising the potential to true Earth of all metalwork (including that of the craft, if connected) to dangerous levels;

(b) inability to establish an equipotential zone external to the craft; and

(c) possible loss of earthing due to long supply cable runs, connecting devices exposed to weather and flexible cable connections liable to mechanical damage.

Particular requirements to reduce the above risks include:

(a) prohibition of the connection of exposed- and extraneous-conductive-parts of the craft to a PME terminal; and

(b) additional protection by 30 mA RCDs in both the craft and the marina installation.

8.4　General requirements

Electrical power installations located at marinas should be installed and the equipment so selected as to minimize the risk of electric shock, fire and explosion. In the design and construction of such works, particular regard should be given to the risk of increased corrosion, movement of structures, mechanical damage, presence of flammable fuel and vapour and the physiological effects of electric shock being increased by a reduction in body resistance and contact of the body with earth potential.

Owing to the harsh working environment of marina installations and potential for abuse and accidental damage by users, particular attention should also be paid to the maintenance and periodic inspection reporting of installations and the general requirements of the Regulations.

8.5　Supplies

Where the supply system is PME, Regulation 9(4) of the ESQCR 2002 prohibits the connection of the neutral to the metalwork of any caravan or boat. While the PME supply may be fed to permanent buildings in the marina, supplies to boats (pleasure craft) must have a separate earth system. A TT system having a separate connection with Earth, independent of the PME earthing system (see Figure 8.3a/b and Figures 7.1 and 7.2), will meet this requirement. Alternatively, protection by electrical separation can be adopted.

This avoids the risks arising from a loss of continuity of the supply PEN conductor.

541.2　The separation of the TT earthing system should be effected at the main distribution board, where the exposed-conductive-parts connected to each system can be more readily identified and inspected periodically. A main earth electrode for the TT system needs to be provided nearby, with no overlap of resistance area with any earthing associated with the PME supply. (See also section 7.5.6 of Chapter 7.)

TN-S supplies may be made available both to permanent shore installations and to pleasure craft.

709.313.1.2　The nominal supply voltage of the installation to pleasure craft should not exceed 230 V single-phase or 400 V three-phase.

Dry dock areas

Some locations may have dry dock areas where construction and maintenance activities can be carried out. The guidance on supplies to these locations can be summarized as follows:

▶ for dry dock areas that form part of a formal workshop type environment under responsible control where construction and maintenance on vessels are carried out, reduced low voltage (110 V centre-tap earthed) and extra-low voltage fixed supplies should be provided. In addition, portable generators could be used that would require the appropriate earthing and protection.

▶ British Waterways have a number of dry dock locations accessible by vessel owners where maintenance activities can be carried out. Here, typical vessel distribution pillar fixed supplies with the appropriate overcurrent and RCD protection would be provided. The vessel owner would connect to this supply and use double insulated electrical equipment connected via the vessel socket-outlets.

▶ British Waterways also have a number of dry dock locations that do not have any fixed installation electrical supply. The supply in this type of location would be via a small generator. This would require the appropriate earthing and protection to be provided.

Note: Persons involved in designing and erecting electrical installations for dry dock facilities should contact the Health and Safety Executive and British Waterways for additional guidance.

8.6 Protection against electric shock

709.410.3 The protective measures of obstacles and placing out of reach (Section 417) shall not be used. Also, the protective measures of non-conducting location (Regulation 418.1) and earth-free local equipotential bonding (Regulation 418.2) shall not be used.

709.411.4 In the UK for a TN system, the final circuits for the supply to pleasure craft or houseboats shall not include a PEN conductor.

Note: In the UK, the ESQCR prohibit the use of a PME system for the supply to a boat or similar construction.

Only permanent onshore buildings may use the electricity distributor's PME earthing terminal. For the boat mooring area of the marina this is not permissible, and entirely separate earthing arrangements must be provided. This is generally achieved by the use of a suitably rated RCD complying with BS EN 61008 with driven earth rods or mats providing a TT system for that part of the installation.

Marina installations are often of sufficient size to warrant the provision of an HV/LV transformer substation. In these, and sometimes in other, circumstances the electricity distributor may be willing to provide a TN-S supply, which is much more suitable for such installations. If the transformer belongs to the marina, a TN-S system should be installed.

8.7 Operational conditions and environmental factors

709.512.2.1.1
709.512.2.1.2 Electrical equipment to be installed on or above jetties, wharves, piers or pontoons must be selected according to the external influences that may be present. Regarding the presence of solid foreign bodies, a minimum degree of protection of IP3X is required, and for presence of water the following applies.

External influence	Minimum index of protection
Presence of water splashes	IPX4
Presence of water jets	IPX5
Presence of waves of water	IPX6

709.512.2.1.3 In the marina environment, particularly at jetties, pontoons etc., consideration must also be given to the possible presence of corrosive or polluting substances.

709.512.2.1.4 Equipment should be located to avoid any foreseeable impact, be provided with local or general mechanical protection and have a degree of protection for external mechanical impact IK08 (see BS EN 62262).

Specific guidance for distribution boards and socket-outlets is given in Section 8.9.

8.8   Wiring systems

709.521.1.5 The following wiring systems should not be used above a jetty, wharf, pier or pontoon:

▶ cables in free air suspended from or incorporating a support wire;
▶ non-sheathed cables in cable management systems, e.g. conduit and trunking;
▶ cables with aluminium conductors; and
▶ mineral insulated cables.

709.521.1.6 Conduit and ducting installations should have suitable apertures or holes and be fixed at an angle sloping away from the horizontal, sufficient to allow for drainage of moisture.

Cables should be selected and installed so that mechanical damage due to tidal and other movement of craft and other floating structures is prevented. To clarify this requirement, cables should be installed in such a manner that they are protected from damage due to:

▶ displacement by movement of craft or other structures;
▶ friction, tension or crushing; and
▶ exposure to adverse temperatures.

See Figure 8.1 for a typical wiring arrangement for offshore pontoons.

At locations where cables are subject to flexing, for example, bridge ramps, between movable jetties and pontoons, flexible cables should be used, such as:

▶ cross-linked insulated flexible cables harmonized type H07RN-F, H07BN4-F or H07RN8-F (insulated and sheathed), e.g. cables to Tables 14,15,16,17 and 20 of BS EN 50525-2-21; and
▶ thermosetting insulated flexible cables harmonized type H07Z-K, e.g. cables to BS EN 50525-3-41 (flexible wiring systems).

Notes:
(a) Cables should be installed in locations where they are protected from physical damage and, wherever practicable, out of water.
(b) Many cable types including PVC insulated and sheathed cables are not suitable for continuous immersion in water. The suitability of the cable types should be checked with the manufacturers. Floating pontoons are usually manufactured with a service void in them, enclosed and accessible from above, to accommodate cables and water piping.

(c) Fixed cables installed permanently under water at a depth of more than 4 m will normally need to be metal sheathed, e.g. lead. Alternatively, a cable that has been specially designed and manufactured for such locations could be used (refer to cable manufacturer for further guidance).

Fixed cables not permanently immersed or at a depth of less than 4 m should be armoured and incorporate extruded MDPE (medium density polyethylene) outer sheath. Note that cables used in this location will not have the same life expectancy as similar cables used on dry land.

Cable type H07 RN8 may be suitable up to 10 m depth of water. However, this type of cable does not offer any mechanical protection against, e.g. boat anchors. Other types of cable (e.g. with mechanical protection) would need to be specially manufactured to the customer's requirements.

543.1.2 **(d)** Due to the possibility of corrosion, the galvanized steel armouring of cables must not be used wholly or in part as a circuit protective conductor (cpc) on the floating section of marinas. A separate protective conductor should be used which, when sized in accordance with Regulation 543.1.2, can be common to several circuits if necessary. The armour must still, however, be connected to protective earth.

(e) Protective bonding connections must be single-core insulated to BS EN 50525-2-31 (H07V-R – PVC type) or BS EN 50525-3-41 (H07Z-R – low smoke, halogen-free type), or BS EN 50525-2-42 (H07G-R – EVA type), or with an oversheath or further mechanical protection as applicable to the particular location.

Table 51 **(f)** Conductor colour coding should be in accordance with the requirements of BS 7671 Table 51. Terminations should be protected against corrosion either by the selection of suitable materials or by covering with grease or water-resistant mastic or paint.

(g) Care should be exercised when installing cables to prevent damage from abrasion due to movement between pontoon sections, etc. Cables must be adequately fixed, protected and supported, and, if necessary, cable types suitable for the flexing movement must be used.

(h) Where cables are installed at onshore locations, due consideration should be given to the routing, depth of lay and protection, especially where heavy traffic and point loads are experienced. Cables should normally be laid above the water table, or cable types suitable for continual immersion used. It is not usually practicable for buried cable duct systems to be made totally watertight. The watertight termination of ducts into drawpits and cable trenches below switchboards is also difficult to achieve.

▶ **Figure 8.1** Typical wiring arrangement from shore to pontoon

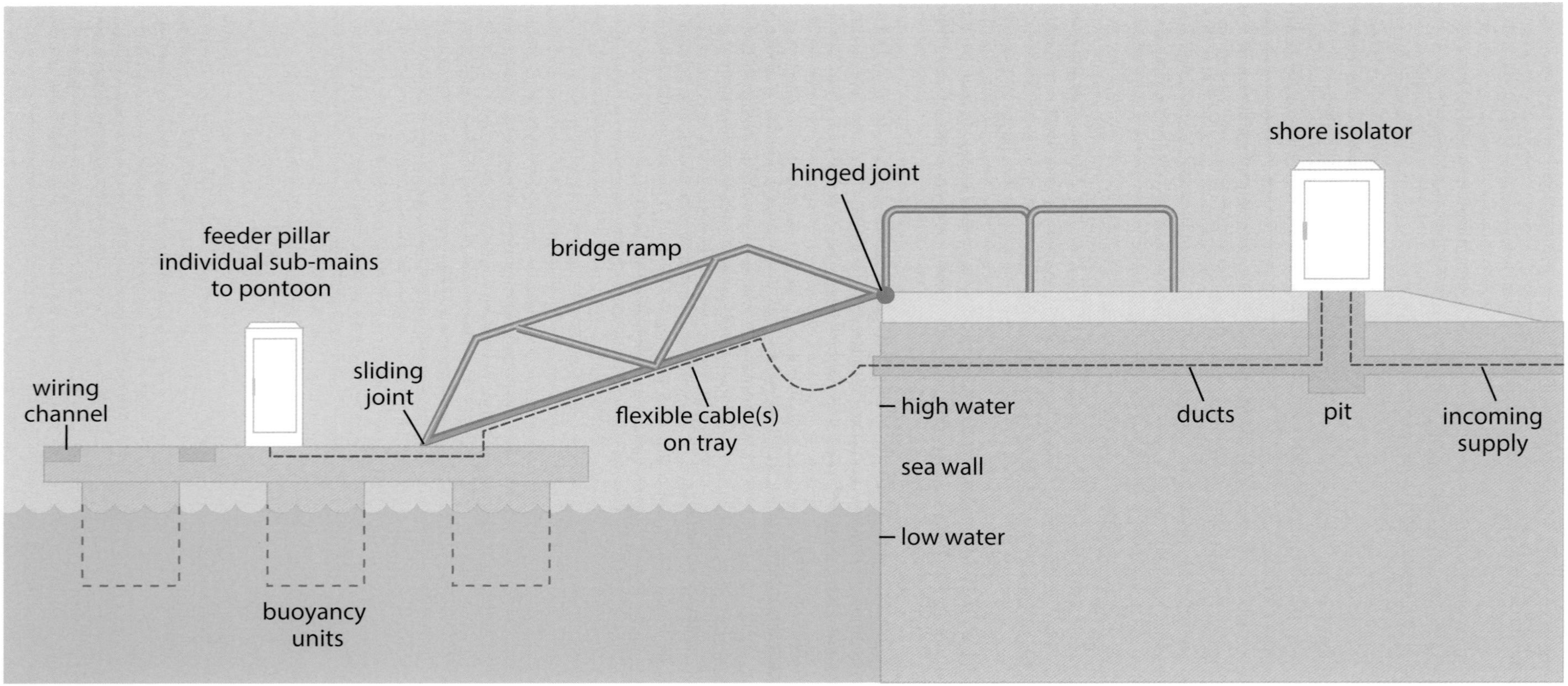

Notes:

(a) Where the particular feeder pillars are in external locations they should be constructed of glass reinforced plastic (GRP), or have GRP housings. GRP is preferred to galvanized steel for protection against corrosion in such environments.

(b) In order to counteract condensation within feeder pillar enclosures, low wattage 'anti-condensation' heaters should be installed.

(c) All feeder pillar and distribution board doors should be fitted with locks to prevent unauthorized access, and have intermediate barriers to protect against accidental contact with live parts when the doors are open. The barriers should provide a degree of protection of at least IPXXB rating or IP2X rating.

8.9 Distribution boards, feeder pillars and socket-outlets

709.531.2 RCBOs must disconnect all poles.

709.553.1.13 Distribution boards and feeder pillars mounted outdoors should meet the degree of protection IP44 rating as a minimum. This will be adequate in sheltered waterways, but the IP code must be selected with reference to the degree of protection necessary for the particular location. The enclosure should be corrosion resistant and give protection against mechanical damage and ingress of dust and sand etc. Distribution boards and feeder pillars supplying marina berths should be sited in the immediate vicinity of berths. When distribution boards and feeder pillars and their associated socket-outlets are mounted on floating installations or jetties, they should be fixed above the walkway and preferably not less than 1 m above the highest water level (see Figure 8.2). This height may be reduced to 300 mm if appropriate additional measures are taken to protect against the effects of splashing (IPX4), but care should be taken to avoid creating a low-level obstacle that may cause risk of tripping on the walkway. When mounted on fixed jetties they should be mounted not less than 1 m above the highest water level.

709.553.1.10
709.553.1.11
709.531.2
709.553.1.8
709.553.1.13

▼ **Figure 8.2** Typical feeder pillar showing socket-outlet height for floating installation

Individual RCD
protection
709.531.2

BS EN 60309-2
up to 63 A
(generally 16 A)
IP 44
709.553.1.8

No more than four sockets at
any one point, one per craft
709.553.1.10 and 11

Not less than 1 m
above highest
water level
709.553.1.13
(may be reduced
to 300 mm if
additional
measures are
taken)

Note: While there is a good argument for socket-outlets on feeder pillars and bollards to be mounted at a high level, they may be at risk from damage from the bows of boats, which can accidentally overshoot the walkways during berthing. A lower mounting level of 300 mm minimum above the walkway can reduce this risk, but care should be taken to avoid creating a low level obstacle, which may trip the unwary.

709.553.1.9 Socket-outlets should be mounted as close as possible to the berth to be supplied and
709.553.1.10 should be installed in a distribution board or separate enclosures. A maximum of four socket-outlets may be grouped together in one enclosure. This is to minimize the

hazard of long trailing flexes. A socket-outlet, either single-phase or polyphase, is only to supply a single pleasure craft or houseboat.

709.553.1.14 Socket-outlet protective conductors shall not be connected to a PME earthing facility.

709.553.1.8 Socket-outlets up to 63 A should be in accordance with BS EN 60309-2 and each outlet should be connected to the circuit protective conductor except where an onshore isolating transformer is used.

709.553.1.12 Generally, socket-outlets with a rating of 16 A should be provided and should have the following characteristics, irrespective of the measure of protection against electric shock:

Single-phase socket-outlets

Rated voltage:	230 V (colour blue)
Rated current:	16 A
Key position:	6 h
Number of poles:	2 plus protective conductor
Construction:	IP44 rating (minimum)

Three-phase socket-outlets

Rated voltage:	400 V (colour red)
Rated current:	16 A
Key position:	6 h
Number of poles:	4 plus protective conductor
Construction:	IP44 rating (minimum)

Note: Whilst 2P + E and 4P + E plugs and sockets are generally used, other configurations may be necessary as in the case of special security circuits indicating unauthorized use of particular socket-outlets on remote monitoring systems.

Where the pleasure craft demand is likely to exceed 16 A, provision should be made for outlets of suitable rating.

709.531 For automatic disconnection of supply, socket-outlets to supply pleasure craft and final circuits intended for fixed supplies to houseboats are to be protected individually by an RCD having the characteristics specified in Regulation 415.1.1. The devices will need to disconnect all poles including the neutral (see Figures 8.3a & b).

709.533 Socket-outlets and fixed connection supplies to houseboats are to be protected by individual overcurrent protective devices, which should comply with the requirements of Chapter 43 of BS 7671 (see Figures 8.3a & b).

709.537.2.1.1 Each distribution cabinet should have at least one means of isolation that will disconnect all live conductors including the neutral. If the device is used to isolate socket-outlets then it should isolate not more than four of these.

Preferably, socket-outlets or groups of single-phase socket-outlets intended for use on the same walkway or jetty should be connected to the same phase. However, individual sockets connected to separate phases of a supply should be located so that they cannot be reached simultaneously. If sockets on separate phases are grouped together on a pillar then a notice warning of the maximum voltage that exists between accessible parts should be provided.

Socket-outlets should include an interlock to prevent the insertion or removal of a plug while under load.

▼ **Figure 8.3** General arrangements for electricity supply to pleasure craft
(a) Connection to mains supply with single-phase socket-outlet

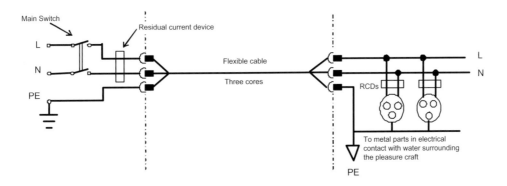

▼ **(b) Connection** to mains supply with three-phase socket-outlet

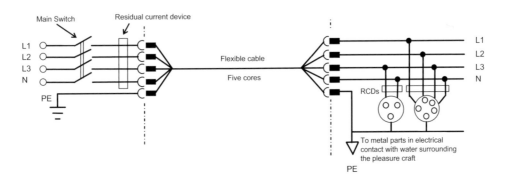

A notice of durable material giving instructions for connection of a pleasure craft to the marina supply is recommended to be placed, where practicable, adjacent to each group of socket-outlets, bearing indelible, weatherproofed and easily legible characters.

Alternatively, the notice should be placed in a prominent position or issued to each berth holder. The notice should contain the text of Figure 8.4.

▼ **Figure 8.4** Instruction notice for connection of supply (Figure 709.3 of BS 7671)

INSTRUCTIONS FOR ELECTRICITY SUPPLY

BERTHING INSTRUCTIONS FOR CONNECTION TO SHORE SUPPLY

This marina provides power for use on your pleasure craft with a direct connection to the shore supply which is connected to earth. Unless you have an isolating transformer fitted on board to isolate the electrical system on your craft from the shore supply system, corrosion through electrolysis could damage your craft or surrounding craft.

ON ARRIVAL

1   Ensure the supply is switched off and disconnect all current-using equipment on the craft, before inserting the craft plug. Connect the flexible cable firstly at the pleasure-craft inlet socket and then at the marina socket-outlet.
2   The supply at this berth is *V, *Hz. The socket-outlet will accommodate a standard marina plug colour * (technically described as BS EN 60309-2, position 6 h).
3   For safety reasons, your craft must not be connected to any other socket-outlet than that allocated to you and the internal wiring on your craft must comply with the appropriate standards.
4   Every effort must be made to prevent the connecting flexible cable from falling into the water if it should become disengaged. For this purpose, securing hooks are provided alongside socket-outlets for anchorage at a loop of tie cord.
5   For safety reasons, only one pleasure craft connecting cable supplying one pleasure craft may be connected to any one socket-outlet.
6   The connecting flexible cable must be in one length, without signs of damage, and not contain joints or other means to increase its length.
7   The entry of moisture and salt into the pleasure craft inlet socket may cause a hazard. Examine carefully and clean the plug and socket before connecting the supply.
8   It is dangerous to attempt repairs or alterations. If any difficulty arises, contact the marina management.

BEFORE LEAVING

1   Ensure that the supply is switched off and disconnect all current-using equipment on the craft, before the connecting cable is disconnected and any tie cord loops are unhooked.
2   The connecting flexible cable should be disconnected firstly from the marina socket-outlet and then from the pleasure craft inlet socket. Any cover that may be provided to protect the inlet from weather should be securely replaced. The connecting flexible cable should be coiled up and stored in a dry location where it will not be damaged.

---

*   Appropriate figures and colours to be inserted:
nominally 230 V 50 Hz blue – single-phase, and
nominally 400 V 50 Hz red – three-phase.

8.10   General notes

8.10.1   Pontoon amenity lighting

It is important that the routes of pontoons and their termination points are clearly delineated.

The lighting may be controlled by either automatic photoelectric cells or time switches, the former being preferred as they sense poor conditions caused by fog, etc. when natural light is waning.

709.512   Luminaires should be of rugged and watertight construction and should preferably be mounted at low level with the light source facing the walkway, not omnidirectional.

8.10.2 Navigation lighting

The local waterway authority should be consulted in order that all necessary and suitably coloured navigation lighting is provided. The light sources should have an extended life expectancy. Photoelectric cell control is preferred to time switches.

8.10.3 Fuelling stations

The relevant local authority should be consulted in order to ensure that the completed installation complies with its requirements. Where applicable, special emergency control facilities should be established onshore. Fuel hoses are required to be non-conducting. Ship/shore bonding cables are not to be used – see *The International Safety Guide for Oil Tankers and Terminals*, 5th Edition.

Electrical equipment in the proximity of fuelling stations should comply with *Guidance for the Design, Construction, Modification, Maintenance and Decommissioning of Filling Stations* jointly published by APEA and Energy Institute (4th edition April 2018).

8.10.4 Metering systems

Metering systems are outside the scope of this Guidance Note and must be agreed between the designer and the marina owner to provide all necessary electricity consumption information for accurate billing. The meters may be required to be installed locally in the feeder pillars for local direct reading, or may be part of a site-wide data network system. The metering system must be fit for the installation and type of use. Functional and safety earthing must be adequate.

Checking the metering for the various main sections of the distribution system may be required in order that the marina operator can use this data in establishing tariffs for the resale of electricity. Such equipment must be installed within the main switchgear and feeder pillars, and must be of adequate rating and quality for the duty required. The increased use of items of electrical equipment exhibiting low power factor characteristics, e.g. dehumidifiers, refrigerators, battery chargers, etc., requires that electricity metering should record suitable data to ensure that the marina operator does not suffer a loss of revenue. (This particularly applies when kVAh metering is installed by the electricity distributor.)

8.10.5 Location of equipment

Due consideration should be given to the location of items of equipment so that they are, as far as practicable, not vulnerable to damage either onshore or offshore at the marina.

In the case of onshore areas there will be the need for clear vehicular movement including large mobile boat hoists, transit lorries and cars, etc. Consequently, the location of feeder pillars and lighting columns requires special attention.

For marina areas, the lighting columns and power supply feeder pillars should be so positioned that the risk of contact with luggage trolleys etc. and such items as the bowsprit of craft is, as far as practicable, reduced to a minimum. This is particularly important where lighting and power supply equipment has moulded enclosures that are unable to withstand such mechanical forces and impact and may, as a result, be damaged.

Site investigations should be carried out at an early stage to determine the maximum wave heights that can be experienced. This is of particular importance at exposed coastal sites. Where marinas have breakwater type pontoons, it is likely that under certain conditions waves will pass over the structure.

8.10.6 Routine maintenance and testing

GN3 Initial inspection and testing of all electrical systems should be carried out on completion of the installation, in accordance with the requirements of Part 6 of BS 7671 and the recommendations of IET Guidance Note 3. A periodic inspection and test of all electrical systems should be carried out annually and the necessary maintenance work implemented. If the site is considered to be exposed, or operational experience shows problems (i.e. misuse), the inspection frequency should be increased to cater for the particular conditions experienced.

All RCDs should be tested regularly by operating the integral test button and periodically by use of a proprietary test instrument to ensure that they conform with the parameters of their relevant product standards, e.g. BS EN 61008.

All tests should be tabulated for record purposes and the necessary forms required by Part 6 of BS 7671 must be provided by the contractor or persons carrying out the inspection and tests to the person ordering the work.

Medical locations 9

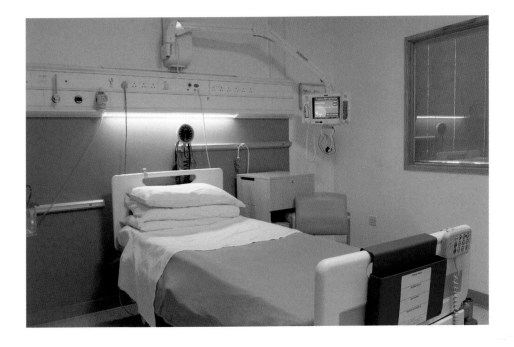

Courtesy of CABLEFLOW INTERNATIONAL LIMITED

9.1 Introduction

Sect 710   Medical locations are incorporated in Part 7 (Special Installations or Locations) of BS 7671:2018. The section is based on the published standard IEC 60364-7-710: 2002, as modified by CENELEC standard HD 60364-7-710: 2012, with additional provisions to satisfy the UK Healthcare requirements.

Any Part 7 document only supplements or modifies the main regulations and should not be read in isolation. Section 710 must be read in conjunction with all relevant regulations stipulated in the main body of BS 7671.

9.2 Scope

710.1   The particular requirements of Section 710 apply to patient healthcare facilities, such as hospitals, private clinics, medical and dental practices, healthcare centres and dedicated medical rooms in the workplace, to provide for the safety of patients and medical staff.

Section 710 also applies to electrical installations in locations designed for medical research only when carried out on patients using medical electrical (ME) equipment. Where applicable, it can also be used in veterinary clinics.

The requirements of this section do not apply to ME equipment or ME systems. This equipment is covered by medical device regulations and generally conforms to the BS EN 60601 series for its design, construction and testing.

Medical supply units (MSU) are covered in BS ISO 11197. This also covers joinery assemblies containing patient care services.

The requirements of other sections of Part 7 may also apply in patient healthcare facilities, for example, mobile or transportable units, locations containing a bath or shower, swimming pools and other basins (such as hydrotherapy pools).

When a change of utilization of the location occurs, it may be necessary to modify the existing electrical installation, in accordance with the current version of BS 7671. This is especially important where the change of use includes life-supporting procedures.

It is important that care is taken so that other installations do not compromise the level of safety provided by installations meeting the requirements of this section. This may be due to:

(a) electromagnetic interference (EMI) and electromagnetic compatibility (EMC) considerations;
(b) effect of fault currents; or
(c) fire.

The Department of Health and Social Care provides supporting information about electrical services supply and distribution in healthcare premises in Health Technical Memorandum HTM 06-01. In Scotland and Wales, this is referred to as SHTM 06-01 and WHTM 06-01, respectively.

Further useful information can be found in the IET publication *Guide to Electrical Installations in Medical Locations*.

9.3 The risks

710.1
710.3
710.4
710.56
Patients undergoing acute care in healthcare establishments (such as hospitals) require enhanced reliability and safety of the electrical installation and safe and reliable operation of the ME equipment used. This is to provide security of supplies and minimize the risk of electric shock.

In medical locations, the risk to patients is increased due to:

▶ the reduction in body resistance, since the skin is often cut or broken or the patient's defensive capacity is either reduced by medication or nullified while anaesthetized; and
▶ the threat from failure of the supply, especially to life-supporting equipment.

The presence of liquids such as blood and saline solutions will also add to the risk, due to increased conductive area and subsequent lower body resistance. A prolonged loss of the mains supply may put the patient's life at risk, especially if the equipment is life-supporting. This equipment must have secure supplies to provide adequate safety.

The use of ME equipment can be split into three main categories:

(a) life support: such as infusion pumps, medical ventilators, anaesthetic equipment and monitors;

(b) diagnostic: such as X-ray imaging, computerized topology (CT) scanners, magnetic resonance imaging (MRI), positron emission tomography (PET) scanners, blood pressure monitors, ultrasound imaging, electroencephalograph (EEG) and electrocardiograph (ECG) equipment; and

(c) treatment: such as interventional X-ray equipment, surgical diathermy, defibrillators, haemodialysis machines and radiotherapy equipment, etc.

Diagnostic equipment such as X-ray, CT and MRI equipment can also be used in the treatment of patients undergoing interventional procedures. Some life-support equipment may also be used for diagnostic purposes. Generally, however, most life-support equipment is utilized within a Group 2 medical location, whereas most diagnostic equipment is utilized within a Group 1 medical location (refer to 9.4 definitions). Those in the treatment category may fall into either, but where a mix of equipment is used, the location is always placed into the more stringent Group 2 category.

9.3.1 Electric shock

Shock hazards due to bodily contact with both DC and AC electric currents are well known and documented in IEC/TR2 60479-1: 2005 *Effects of current on human beings and livestock – Part 1: General aspects*. This Technical Report indicates that, depending on skin resistance, path and area of contact, impedance of the human body, environmental conditions and duration, currents of the order of 10 mA passing through the human body can result in muscular paralysis, followed by respiratory paralysis. Eventual ventricular fibrillation can occur at currents just exceeding 20 mA.

In a normal environment, the average body impedance may be assumed to be 2 kΩ, but in a medical location, where ME equipment is being used, the value given in BS EN 60601-1 is only 1 kΩ. Therefore, the risk in medical locations is increased. This is why the conventional touch voltage is reduced from 50 V AC to 25 V AC in medical locations of Groups 1 and 2.

This reduction in average body resistance is due to the natural protection of the human body being considerably reduced when certain clinical procedures are being performed. For example, patients undergoing treatment may experience a reduction in their body resistance due to their skin being broken; due to the presence of liquids such as blood and saline solutions; or due to their defensive capacity being either reduced by medication or nullified while anaesthetized.

Currents, flowing via applied parts of ME equipment, when introduced directly to the heart[1] can interfere with cardiac function at levels considered safe under other circumstances. In order to protect the patient, the requirements of such equipment are enhanced.

Patient leakage current[2] that can flow via a patient to earth is normally greatest when the equipment earth is disconnected (single fault condition). ME equipment product standards (BS EN 60601-1) set limits to the amount of leakage current that can flow in the patient circuit, both when the protective earth conductor is connected or disconnected. Patient leakage currents of the order of 10 μA have a probability of 0.2 % of causing ventricular fibrillation or pump failure when applied through a small area of the heart. At 50 μA, the probability of ventricular fibrillation increases to the order of 1 % (BS EN 60601-1).

9.3.2   Medical electrical equipment

Although outside the scope of BS 7671, it is useful to have some basic understanding of ME equipment.

The BS EN 60601 series of standards specify the allowable limits of leakage currents produced by ME equipment to a required level to provide for patient safety. This is why only ME equipment meeting the standards should be used clinically within the patient environment. It should also be noted that the allowable levels of patient leakage current depend on the classification of the applied parts, which are split into three categories: B, BF and CF.

(a) **Type B** (Body) applied part: these parts are either connected to earth or referenced to earth. Their allowable patient leakage current is 100 μA in Normal Condition (NC) and 500 μA in Single Fault Condition (SFC).

(b) **Type BF** (Body Floating) applied parts: these parts are isolated from earth and from other parts of the equipment. They offer an enhanced degree of protection over Type B. Their allowable patient leakage current is 100 μA in NC and 500 μA in SFC or 5000 μA with mains on applied parts.

(c) **Type CF** (Cardiac Floating) applied parts: these parts are isolated from earth and from other parts of the equipment to a higher degree than BF. They are deployed inside (intracardiac) [1] or near the heart. Their allowable patient leakage current is 10 μA in NC and 50 μA in SFC, including mains on applied part.

ME equipment will also limit the allowable touch leakage current[3] for items such as the enclosure to 100 μA in NC and 500 μA in SFC.

---

1   A procedure whereby an electrical conductor is placed within the heart of a patient or is likely to come into contact with the heart, such conductor being accessible outside the patient's body. In this context, an electrical conductor includes insulated wires such as cardiac pacing electrodes or intracardiac ECG electrodes, or insulated tubes filled with conducting fluids (catheter) connected to ME equipment.

2   Patient leakage current: current flowing from the patient connections via the patient to earth, or originating from the unintended appearance of a voltage from an external source on the part to earth patient and flowing from the patient via the patient connections of an F-type applied part to earth.

3   Touch leakage current: leakage current flowing from the enclosure or from parts thereof, excluding patient connections, accessible to any operator or patient in normal use, through an external path other than the protective earth conductor, to earth or to another part of the enclosure.

Note: Where the requirements of Section 710 are properly fulfilled, the risk of a dangerous touch leakage current occurring inadvertently, from medical staff simultaneously touching an intracardiac conductor and the electrical installation earth (e.g. via the enclosure of ME equipment), are negligible.

Note: Further information on the modelling of electric shock to the human body can be found in the IET publication *Guide to Electrical Installations in Medical Locations*.

9.3.3 Security of supply

Additional to the consideration of risk from electric shock, some ME equipment and ME systems (life-support equipment and surgical equipment) perform such vital functions that prolonged loss of supply would pose an unacceptable risk to patients.

This has implications not only for the provision of safety (emergency) power supplies, but can also render some conventional protective measures unsuitable. Hence, for example, when protecting circuits supplying critical medical equipment, restrictions are stipulated on the use of RCDs.

9.4 Definitions

Part 2: BS 7671 It is important that the definitions used in Section 710 are clearly understood. They are reproduced below.

9.4.1 Medical location

Location intended for purposes of diagnosis, treatment including cosmetic treatment, monitoring and care of patients.

9.4.2 Patient

Living being (person or animal) undergoing a medical, surgical or dental procedure. A person undergoing surgical treatment for cosmetic purposes may also be regarded as a patient.

9.4.3 Medical electrical (ME) equipment

Electrical equipment having an applied part or transferring energy to or from the patient or detecting such energy transfer to or from the patient and which is:

(a) provided with not more than one connection to a particular supply mains; and
(b) intended by the manufacturer to be used:
    (i) in the diagnosis, treatment or monitoring of a patient; or
    (ii) for compensation or alleviation of disease, injury or disability.

Note: ME equipment includes those accessories as defined by the manufacturer that are necessary to enable the normal use of the ME equipment.

Note: This equipment should meet the requirements of the BS EN 60601-1 series of standards.

9.4.4 Applied part

Part of ME equipment that in normal use necessarily comes into physical contact with the patient in order for the ME equipment or an ME system to perform its function.

9.4.5 Group 0

Medical location where no applied parts are intended to be used and where discontinuity (failure) of the supply cannot cause danger to life.

9.4.6 Group 1

Medical location where discontinuity (failure) of the supply does not represent a threat to the safety of the patient and applied parts are intended to be used:

(a) externally; or
(b) invasively to any part of the body, except where Group 2 applies.

9.4.7 Group 2

Medical location where applied parts are intended to be used, and where discontinuity (failure) of the supply can cause danger to life, in applications such as:

(a) intracardiac procedures; and
(b) vital treatment and surgical operations.

9.4.8 Medical electrical (ME) system

Combination, as specified by the manufacturer, of items of equipment, at least one of which is ME equipment to be interconnected by functional connection or by use of a multiple socket-outlet.

Note: The system includes those accessories that are needed for operating the system and are specified by the manufacturer.

9.4.9 Patient environment

Any volume in which intentional or unintentional contact can occur between a patient and parts of the ME equipment or ME system or between a patient and other persons touching parts of the ME equipment or ME system. For illustration, see Figure 9.1.

Fig 710.1   ▼ **Figure 9.1** Patient environment (BS EN 60601-1:2006)

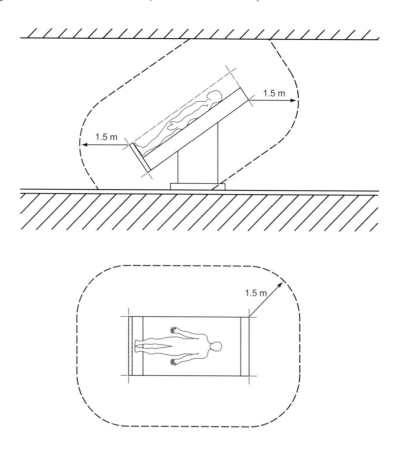

Note: The dimensions in the figure show the minimum extent of the patient environment in a free surrounding. It applies when the patient's position is predetermined; if it is not, all reasonably practicable patient positions should be considered.

9.4.10 Medical IT system

An IT electrical system fulfilling specific requirements for medical applications.

Note: The combination of a specified IT transformer and a specified insulation monitoring device (IMD) is referred to as a medical IT system.

Note: These supplies are also known as isolated power supply systems (IPS).

Note: Currently, many medical IT systems are backed up by uninterruptible power supplies (UPS). Refer to Figure 9.4 for a typical outline arrangement.

9.5 Assessment of general characteristics

710.3 In order to determine the appropriate (safety) classification and group number of a medical location, it is necessary that the relevant medical staff indicate which medical procedures will take place within the location and the effect supply loss would have on patient safety. This may also include consideration of the effects of having to repeat a procedure, as well as the ancillary support services required for business continuity aspects.

Based on the intended use and assessment of the medical procedures being performed, the appropriate classification for the location can be determined (refer to 9.22).

Note: Further information can be found in the IET publication *Guide to Electrical Installations in Medical Locations*.

The allocation of a group number and classification of safety services for medical locations shown in Table 9.1 should only be used as a **guide**, not a substitute for proper fact-finding.

Where a medical location may be used for different medical procedures, the requirements of the higher group classification should be applied (guidance is shown in Table 9.1). Where a change of utilization takes place, it is important to perform a new assessment, in case the area is now a higher group rating.

▼ **Table 9.1** Example allocation of Group numbers & classification for safety services of medical locations

	Medical location	Group			Classification (refer to 9.22)	
		0	1	2	≤ 0.5 s	> 0.5 s ≤15 s
1	Massage room	X	X			X
2	Bedrooms		X			X
3	Delivery room		X		X^a	X
4	ECG, EEG, EHG room		X			X
5	Endoscopic room		X^b		X	X^b
6	Examination or treatment room		X		X	X
7	Urology room		X^b		X	X^b

Medical location	Group			Classification (refer to 9.22)	
	0	1	2	≤ 0.5 s	> 0.5 s ≤15 s
8 Radiological diagnostic and therapy room		X	X	X	X
9 Hydrotherapy room		X			X
10 Physiotherapy room		X			X
11 Anaesthetic area			X	Xa	X
12 Operating theatre			X	Xa	X
13 Operating preparation room			X	Xa	X
14 Operating plaster room			X	Xa	X
15 Operating recovery room			X	Xa	X
16 Heart catheterization room			X	Xa	X
17 Intensive care room			X	Xa	X
18 Angiographic examination room			X	Xa	X
19 Haemodialysis room		X			X
20 Magnetic resonance imaging (MRI) room		X	X	X	X
21 Nuclear medicine		X			X
22 Premature baby room (NICU/SCBU)			X	Xa	X
23 Intermediate Care Unit (IMCU)			X	X	X

Notes on Table 9.1:

(a) Specific luminaires, such as operating or procedure lights, that require a power supply within 0.5 s, and life-supporting ME equipment that requires a power supply within 0.5 s.

(b) Not being an operating theatre.

A definitive list of medical locations showing their assigned groups is impracticable, as is a list of the use to which locations (rooms) might be put. Table 9.1 is **only a guide** for designers and stakeholders to consider typical locations. Any designation of group/classification based on risk assessment and discussed locally by the designer with the persons responsible for clinical procedure or safety (for example, an 'electrical safety group') is also acceptable, provided that all relevant risk associated with the installation is mitigated by an appropriate measure. The requirements of Regulation 710.3 **cannot** be satisfied by using this Table alone. The final outcome can only be determined by following the procedure identified in Regulation 710.3.

Designers and stakeholders have to ascertain the suitability of any electrical equipment brought into the patient environment and whether it is connected to an IT, TN or TT system, where TN and TT systems should be RCD-protected. For example, mobile X-ray equipment will require a TN supply even when used in a Group 2 location.

Special consideration should be given when current-carrying equipment possessing non-specified leakage currents is permanently installed in a Group 2 medical location.

The primary difference between the three medical location groups is the effect a loss of supply has on safety. The reference to applied parts in the definitions is only to indicate the use of ME equipment and ME systems. Group 2 has the highest risk, due to procedures where the loss of supply to life-supporting equipment would endanger life.

9.6 Purposes, supplies and structure

9.6.1 Types of system earthing

710.312.2 Protective earth and neutral (PEN) conductors should not be used in medical locations and medical buildings downstream of the main distribution board.

> **Note:** In Great Britain, Regulation 8(4) of the ESQCR prohibits the use of PEN conductors in a consumer's installation.

9.7 Supplies

9.7.1 General

710.313.1 In medical locations, the distribution system should be designed and installed to facilitate the automatic changeover from the main distribution network to the electrical safety source feeding essential loads, as required by Regulation 560.5.

9.8 Protection against electric shock

9.8.1 Basic protective provisions

710.410.3.5 The protective measures of obstacles and placing out of reach (Regulation 417.1 of BS 7671) shall not be used.

710.410.3.6 The protective measures of a non-conducting location (Regulation 418.1 of BS 7671), earth-free local equipotential bonding (Regulation 418.2 of BS 7671) or electrical separation for the supply of more than one item of current-using equipment (Regulation 418.3 of BS 7671) shall not be used.

Only protection by insulation of live parts or by the use of Class II equipment are permitted.

It is noted that a medical IT system does not use electrical separation as the sole means of protection against electric shock (Regulation 710.411.6).

9.9 Requirements for fault protection (protection against indirect contact)

9.9.1 Automatic disconnection in case of a fault

710.411.3.2.1 Some equipment, such as switched-mode power supply equipment, and some ME equipment, such as fixed and portable X-ray units, may have high protective conductor leakage currents. The increased use of Information Technology equipment can also lead to increased total protective conductor currents.

It is important that care is taken to avoid unwanted tripping of the RCD by simultaneous use of many items of equipment connected to the same circuit. Tripping could affect patient safety or the ability to carry on treatment or diagnosis.

Consideration of the use of multiple circuits to enable equipment to utilize another circuit in the event of an unwanted trip, as well as dedicated circuits for certain equipment, will help to mitigate this risk.

The correct type of RCD for Group 1 and Group 2 medical locations should be selected. This can either be Type A according to BS EN 61008 and BS EN 61009, or Type B according to BS EN 62423, depending on the possible fault current arising.

Type AC RCDs provide tripping for residual sinusoidal alternating currents only. These devices may not therefore detect all leakage currents that could affect patient safety and are not permitted.

Type A RCDs provide tripping for:

(a) residual sinusoidal alternating currents;
(b) residual pulsating direct currents; and
(c) residual pulsating direct currents superimposed by a smooth direct current of 6 mA, with or without phase-angle control, independent of the polarity.

Type B RCDs provide tripping as for Type A and for:

(a) residual sinusoidal currents superimposed by a pure direct current;
(b) residual sinusoidal currents up to 1000 Hz;
(c) pulsating direct currents superimposed by a pure direct current; and
(d) residual currents resulting from various configurations of rectifier circuits.

For example, a typical 3-phase X-ray generator uses a rectifier circuit to generate a DC intermediate voltage. Because of this, equipment manufacturers often advise the use of a Type B RCD.

710.411.3.2.5 To limit the maximum voltage present under a fault condition in Groups 1 and 2 medical locations, the voltage presented between simultaneously accessible exposed-conductive-parts and/or extraneous-conductive-parts should not exceed 25 V AC or 60 V DC.

9.9.2 TN systems

710.411.4 Regulation 710.411.4 states that RCDs having the characteristics specified in Regulation 415.1.1 of BS 7671 shall be used:

(a) in final circuits of Group 1 rated 32 A and below; and
(b) in final circuits of Group 2 (including general lighting circuits), except those of the medical IT system specified in Regulation 710.411.6.

Note: BS 7671 does not require general lighting circuits to be connected to the medical IT system. This recommendation is repeated in HTM 06-01 - 2017.

This clarifies the requirements for RCD protection. It is also permissible to use RCDs in circuits above 32 A in Group 1 locations, if appropriate.

For information, Regulation 415.1.1 of BS 7671 states that the use of RCDs with a rated residual operating current not exceeding 30 mA is recognized in AC systems as additional protection in the event of failure of the provision for basic protection and/or the provision for fault protection or carelessness by users.

9.9.3 TT systems

710.411.5 In medical locations of Group 1 and Group 2, RCDs shall be used as protective devices except for circuits of a medical IT system specified in Regulation 710.411.6.

9.9.4 Medical IT systems

710.411.6 Regulation 710.411.6 reflects the changes to Regulation 710.411.4, stating the following:

In Group 2 medical locations an IT system, including the measures of Regulations 710.512.1.1, 710.411.6.3.1 and 710.411.6.3.2, shall be used for final circuits of ME equipment and ME systems intended for life-support and surgical applications within the patient environment, excluding:

(a) equipment with a rated power greater than 5 kVA;
(b) X-ray equipment; and
(c) the supply of movements of fixed operating tables.

For each group of rooms serving the same function, at least one medical IT system is necessary.

The list of circuits in (a) to (c) above is not exhaustive.

A system constructed to the requirements of Regulation 710.411.6 is known as a medical IT system.

Note: Any non-medical electrical equipment located or brought into the patient environment has to be 'risk-assessed' for its suitability for use in this environment in accordance with BS EN 60601-1.

The medical IT system incorporates a number of features to enhance the supply continuity in the event of a first fault to earth. The isolating transformer (Regulation 710.512.1.1) prevents the loss of supply of all connected equipment when a first fault to earth occurs on a single item.

The design and specification of the transformer (see 9.15.1) maintains the leakage current of the affected device to the allowable limits for ME equipment in a single fault condition (500 µA). The system also alerts users to this fault condition by incorporating an IMD (Regulation 710.411.6.3.1), so that they may take action to remove the faulty equipment before a second fault occurs.

The transformer is also monitored (Regulation 710.411.6.3.2) for overload and high temperature and will alert the user to take action to reduce the load before any loss of supply occurs.

All these measures will provide a continuous supply to the connected equipment and form the medical IT system.

The list of excluded equipment includes mobile imaging and other high-power treatment and diagnostic equipment brought into the patient environment. This is to prevent overload of the medical IT transformer due to high peak and/or continuous current demands.

For illustration of a typical theatre layout, refer to Figure 9.2.

▼ **Figure 9.2** Typical theatre layout of medical IT system with insulation monitoring

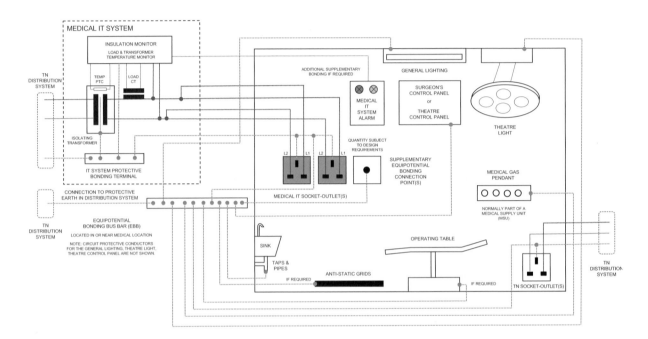

Note: Regulation 710.415.2.3 requires the equipotential bonding busbar (EBB) to be connected to the system earthing using a protective conductor having a cross-sectional area greater than or equal to the largest cross-sectional area of any conductor connected to the EBB.

Note: The medical IT transformer supplying final circuits in Group 2 locations should be designed to BS EN 61558-2-15, providing galvanic isolation between primary and secondary winding.

The transformers are generally defined as single-phase, 1:1 transformers rated at a nominal 230 V when supplying final circuits of medical IT socket-outlets or single-phase permanently connected equipment in Group 2 locations. The output is isolated from earth; nominally, each leg has a potential of approximately 115 V to earth (capacitive coupling). A neutral conductor does not exist.

Regulation 710.512.1.1(iii) states: "If the supply of three-phase loads via a medical IT system is also required, a separate three-phase transformer shall be provided for this purpose". This would imply that a three-phase medical IT transformer may only be dedicated to a specific three-phase load, i.e. not for distribution to single-phase socket-outlets or single-phase permanently connected equipment.

In the UK there is no recognized regulation defining the colour coding of the output (secondary) conductors of a single-phase medical IT transformer.

However, by convention, some UK designers use the colours brown-brown (or brown sleeves if composite cables are used) to identify the secondary conductors, marking them L1 and L2. Generally, L1 is connected to the 'L' terminal and L2 to the neutral terminal of the medical IT socket-outlet. This should not be confused with a 3-phase transformer output where Table 51 of BS 7671 identifies the output colours as Brown, Black and Grey (L1, L2 and L3).

With regard to the colour coding of the final circuits of a medical IT system, the IET publication *Guide to Electrical Installations in Medical Locations* states the following:

As there is no connection to earth for either conductor, they are referred to as L1 and L2, i.e. neither is referred to as the phase or the neutral. However, by the requirements of HTM 06-01 these cables are required to be both brown and labelled L1 and L2, with L2 being connected into the neutral (terminal) of the Medical IT socket-outlet. This convention is acceptable, neither cable is a neutral conductor, although not expressly stated in BS 7671.

The same publication shows this convention in Figure 20.8.

The HTM 06-01 referred to in the above IET publication is the 2007 edition.

Similarly, HTM 06-01 (published in April 2017) states:

Table 51 in BS 7671 identifies conductors as brown–blue. When the conductors forming a Medical IT circuit are coloured brown–brown, they should be identified as L1 and L2 at the points of termination, or in composite cables of brown–blue, the blue conductor should be sleeved brown and labelled L2.

710.411.6.3.1 This regulation covers the insulation monitoring requirements.

For each group of rooms serving the same function, at least one medical IT system is necessary. The IT system shall be equipped with an insulation monitoring device (MED-IMD) in accordance with Annex A and Annex B of BS EN 61557-8.

For each medical IT system, an audible and visual alarm system incorporating the following components shall be provided, so that it can be permanently monitored by the medical staff and any alarm reported to the technical staff:

(a) a green signal lamp to indicate normal operation.
(b) a yellow signal lamp that lights when the minimum value set for the insulation resistance is reached. It shall not be possible for this light to be cancelled or disconnected.
(c) an audible alarm that sounds when the minimum value set for the insulation resistance is reached. This audible alarm may be silenced.
(d) the yellow signal shall extinguish on removal of the fault and when the normal condition is restored.

Documentation shall be easily readable in the medical location and it shall include:

(a) the meaning of each type of signal; and
(b) the procedure to be followed in case of an alarm at first fault.

The indicator colours of the visual alarm system associated with the medical IT system align with the colour of indicator lights for medical equipment given in BS EN 60601-1. Care should be taken over the selection of any indicator lights visible within the patient environment to avoid any confusion by medical staff.

710.411.6.3.2 Monitoring of overload and high temperature for the medical IT transformer is required to avoid unnecessary tripping under these conditions. The alarm is raised when the load current exceeds the rated output of the transformer. However, if it is within the specification of the equipment to adjust the set point, it is desirable that the alarm is raised earlier, for example, at 10 % below the rated output.

710.411.6.3.3 In addition to an IMD, consideration should be given to the installation of fault location systems in order to localize insulation faults in any part of the medical IT system. The insulation fault location system should be in accordance with BS EN 61557-9.

For illustration of a typical medical IT system arrangement, see Figure 9.3.

▼ **Figure 9.3** Typical medical IT system arrangement

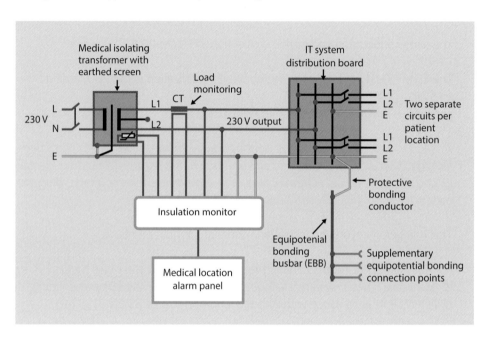

Although Regulation 710.553.1 stipulates a minimum of two final circuits to supply socket-outlets in the medical IT system, the number of final circuits can be increased, if required, to meet anticipated equipment use and loading considerations.

710.415.1 Regulation 710.415.1 states: "Where a medical IT system is used, additional protection by means of an RCD is not required". It should **not** be employed for first fault protection.

The medical IT system provides first fault protection in conjunction with the IMD (MED-IMD).

9.9.5   Functional extra-low voltage (FELV)

710.411.7 In medical locations, functional extra-low voltage (FELV) shall not be used as a method of protection against electric shock, due to safety concerns.

9.10   Protection by SELV and PELV

710.414.1 When using SELV and/or PELV circuits in Group 1 and Group 2 medical locations, the nominal voltage applied to current-using equipment shall not exceed 25 V AC rms or 60 V ripple-free DC. This reduction from the general requirements is referred to in the section about risks (9.3). Protection by basic insulation of live parts, as required by Regulation 416.1 or by barriers or enclosures as required by Regulation 416.2, shall be provided – these are as indicated by the basic protective measures in section 9.8.1.

710.414.4.1 In Group 2 medical locations, where PELV is used, exposed-conductive-parts of equipment, such as operating theatre luminaires, shall be connected to the circuit protective conductor. This in turn is connected to the EBB by supplementary bonding conductors.

9.11 Supplementary protective equipotential bonding (additional protection)

710.415.2.1 In each Group 1 and Group 2 medical location, supplementary equipotential bonding is required. The supplementary bonding conductors are connected to the EBB for the purpose of equalizing potential differences between the following parts, which are located, or may be moved into, the 'patient environment':

(a) circuit protective conductors;
(b) extraneous-conductive-parts;
(c) electric field interference screening, if installed;
(d) conductive floor grids, if installed; and
(e) metal screen of isolating transformers, via the shortest route to the earthing conductor.

Supplementary equipotential bonding connection points, which allow for the connection of ME equipment, shall be available in Group 2 and should also be considered in Group 1 medical locations.

The designer, in consultation with the end-user (for example, the 'electrical safety group'), should determine the appropriate number of supplementary equipotential bonding connection points.

Regulation 710.415.2.1 recommends the following arrangements:

(a) for Group 1 medical locations: a minimum of one supplementary equipotential bonding connection point to be provided per patient location; and
(b) for Group 2 medical locations: a minimum number of four supplementary equipotential bonding connection points, but not less than 25 % of the total number of individual medical IT socket-outlets, to be provided per patient location.

It is expected that the connection points should be suitable for use with the intended connection leads.

Manufacturers of fixed conductive non-electrical patient supports, such as operating theatre tables, physiotherapy couches and dental chairs, may require the equipment to be connected to the equipotential bonding conductor, if such a connection point is available on the equipment. It is important to follow the manufacturers' instructions. Modification of the equipment to create such a connection point is not allowed.

9.12 Resistance of protective conductors in Group 1 and Group 2 medical locations

710.415.2.2 In Group 1 and Group 2 medical locations, the resistance of the protective conductors, between the earth terminal of any socket-outlet (or fixed equipment) and any exposed-conductive-part and/or extraneous-conductive-part, is required to be such that the voltages given in Regulation 710.411.3.2.5 are not exceeded and, subject to that limitation, the measured resistance between the earth terminal of any socket-outlet (or fixed equipment) and any extraneous-conductive-part shall not exceed 0.2 Ω.

It should be noted that in many instances the required value will be less than 0.2 Ω.

The required value, subject to protective device characteristics, serves to limit any potential difference between the exposed-conductive-parts of any ME equipment and other parts of the electrical installation which may be touched, such that the potential difference should remain within the limits of Regulation 710.411.3.2.5 under fault conditions.

9.12.1 Specification of the equipotential bonding busbar (EBB)

710.415.2.3 The EBB is required to be located in or near the medical location, to maintain the shortest possible connection paths, using a radial pattern to avoid 'earth loops' that may exacerbate electromagnetic disturbances and short lengths to maintain the required resistance values. In addition, the EBB should be connected as indicated in Regulation 710.415.2.3:

> *The equipotential bonding busbar shall be connected to the system earthing using a protective conductor having a cross-sectional area greater than or equal to the largest cross-sectional area of any conductor connected to the equipotential bonding busbar.*

This requirement aligns the regulation with the technical intent of HD 60364-7-710. To enable testing and identification, the connections need to be arranged so that they are accessible, can be individually disconnected and are clearly and visibly labelled.

Where an IT system protective bonding terminal, such as that found within a medical IT transformer panel, is located some distance away, the connection between the EBB and this panel is by a larger size protective conductor.

It is recommended as general guidance that:

(a) the EBB is located in or near the location;
(b) the EBB is easily found (visible or with suitable marking to the location), with adequate access to carry out any maintenance and periodic inspection/testing, etc.;
(c) the EBB enclosure should be permanently labelled for the location and the group number;
(d) record drawings should be marked with the location of the EBB; and
(e) for safety, all related circuits must be isolated before any of these conductors are disconnected.

9.13 Arc fault detection devices (AFDD)

710.421.1.201 Arc fault detection devices (AFDDs) are devices designed to detect high energy arcs
421.1.7 within cables. They will not detect the low energy switching arcs that are normally encountered. Regulation 421.1.7 has been introduced in BS 7671:2018, recommending the installation of AFDDs conforming to BS EN 62606 to mitigate the risk of fire caused by arc faults in AC final circuits.

Examples of where these devices can be used are listed in Regulation 421.1.7. These include locations where combustible materials or structures are present, where there are fire-propagating structures and for irreplaceable goods. The locations also include premises with 'sleeping accommodation'.

It should be noted that sleeping accommodation within Groups 1 and 2 of a healthcare premises is always under supervision. Therefore, in Groups 1 and 2 medical locations, the installation of AFDDs is not required, although these could be used in medical locations of Group 0, subject to a risk assessment.

9.14 Measures against electromagnetic disturbances

710.444 Consideration has to be taken of electromagnetic interference (EMI) and electromagnetic compatibility (EMC), where medical electrical equipment is used. Further information can be found in the instructive narrative of Section 444 of BS 7671:2018, as well as in HTM 06-01.

9.15 Operational conditions

9.15.1 Transformers for medical IT systems

710.512.1.1 It is a requirement for safety that transformers used for medical IT systems shall comply with BS EN 61558-2-15, to ensure they meet the required standard for maximum allowable leakage currents. They must be installed in close proximity to the medical location. This proximity is not only a cable capacitive coupling issue, but one of security of supply, for example, in the event of a fire or cable fault.

The specification of the transformer includes the following:

(a) the leakage current of the output winding to earth and the leakage current of the enclosure, when measured in no-load condition and with the transformer supplied at rated voltage and rated frequency, shall not exceed 0.5mA (500 µA);

(b) at least one single-phase transformer per room or functional group of rooms is to be used to form the medical IT systems for mobile and fixed equipment. The rated output shall be not less than 0.5 kVA and shall not exceed 10 kVA. Where several transformers are needed to supply equipment in one room, they shall not be connected in parallel;

(c) if the supply of three-phase loads via a medical IT system is also required, a separate three-phase transformer shall be provided for this purpose;

(d) capacitors shall not be used in transformers for medical IT systems.

Note: For monitoring, see Regulation 710.411.6.3.1.

The conductors constituting the output of the secondary of the medical IT transformer (Figure 9.3) are not directly referenced to earth. However, a potential of approximately half the rated output voltage will be measured, when using a high impedance voltmeter, with respect to earth through capacitive coupling.

When selecting circuit-breakers for transformer input and output circuits, designers need to take into account possible high inrush currents associated with both the medical IT transformer and, in particular, inrush currents associated with equipment connected to final circuits on the output of the medical IT transformer. In this instance, Regulation 533.2.1 of BS 7671 applies.

533.2.1 This regulation calls for consideration of peak load current values, rather than just steady-state values.

BS EN 61558-2-15 states that the maximum inrush current shall not exceed 8 times the peak value of the rated input current. However, it is permitted to increase this value to 12 times the rated input current, provided the latter value is reflected on the transformer marking.

Neither CENELEC HD 60364-7-710 (2012) nor BS 7671:2018 Section 710 "Medical Locations" refer to the use of uninterruptible power supply (UPS) equipment as safety back-up to the medical IT system in Group 2 medical locations, where the latter serves life-support equipment. The required safety has been generally provided by standby generators and embedded equipment batteries. However, static UPS units are being used frequently to provide such safety back-up.

Additional information on UPS system back-up can be found in HTM 06-01.

Refer to Figure 9.4 for a typical outline arrangement of a UPS supply.

▼ **Figure 9.4** Typical arrangement of a UPS back-up to a medical IT system

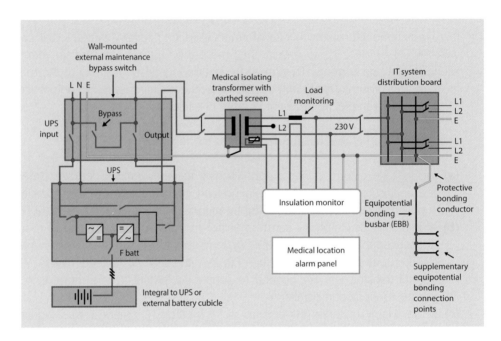

Note: The above arrangement represents a typical outline. It does not include any added details of all the controls/switchgear associated with final designs.

9.15.2 Power supply for medical locations of Group 2

710.512.1.2 Regulation 710.512.1.2 states that in case of a fault or a failure of supply, a total loss of power in a Group 2 medical location shall be prevented.

This does not solely refer to medical IT supplies, but applies equally to the level of resilience a designer must apply to the distribution circuits supplying a Group 2 medical location. For example, the loss of one distribution circuit through fault or fire must not cause total loss of supply to that location. The supplies to these distribution and final circuits may be subject to the requirements of Chapter 56 of BS 7671.

9.15.3 Distribution boards

710.511.1 Distribution boards are required to meet the requirements of BS EN 61439 series. Those for Group 2 medical locations must be installed in close proximity to the areas they serve and be clearly labelled.

9.15.4 Explosion risk

710.512.2.1 To minimize the risk of ignition of flammable gases, which are no longer in general use in the UK, mains-supplied electrical devices, such as socket-outlets and switches, are required to be located at a distance of at least 0.2 m from the centre of the medical gas outlet to the first point of incursion with the socket-outlet moulding outlet (BS ISO 11197).

Where oxygen is installed within medical supply units or walls, this may present a potential risk of flammability through an oxygen-enriched atmosphere where a failure of the medical gas terminal unit or pipeline is undetected (single fault condition) and must be considered. Where medical gases and electrical services are located in the same enclosure and intended to supply patient locations, they will be considered as a medical supply unit according to BS ISO 11197. Guidance on medical supply units is also given in HTM 08-03, HTM 02-01 and HTM 06-01.

9.15.5 Diagrams and documentation

710.514.9.1 It is required to provide plans of the electrical installation, together with records, drawings, wiring diagrams and modifications relating to the medical location. Information may include, but is not limited to:

(a) single-line overview diagrams showing the distribution system of the normal power supply and power supply for safety services in a single-line representation;
(b) distribution board block diagrams showing switchgear and controlgear and distribution boards in a single-line representation;
(c) schematic diagrams of controls;
(d) the verification of compliance with the requirements of Regulation 710.4 in BS 7671:2018 and
(e) functional description for the operation of the safety power supply services and of the safety power supply system.

9.16 Selection and erection of wiring systems in Group 2 medical locations

710.52 Any wiring system within Group 2 medical locations shall be exclusively for the use of equipment and accessories within those locations.

This is required so that any faults and/or electromagnetic disturbances occurring in the wiring serving other non-medical locations and not complying with the requirements for a medical location do not impair patient safety.

9.17 Overcurrent protective devices: protection of wiring systems in Group 2 medical locations

710.531.2 Although overcurrent protective devices (such as fuses) may be used in the primary circuit of the transformer, for short-circuit protection only, overload current protection shall not be used in either the primary or secondary circuit of the transformer of a medical IT system.

However, overcurrent protection against overload and short-circuit currents is still required for each final circuit.

Figure 9.3 shows a typical medical IT system arrangement.

9.18 Socket-outlets protected by RCDs

710.531.3.2 Regulation 710.531.3.2 requires that for each circuit protected by an RCD having the characteristics specified in Regulation 415.1.1, consideration must be given to reducing the possibility of unwanted tripping of the RCD due to excessive protective conductor currents produced by equipment during normal operation.

As some equipment, including ME equipment such as fixed and portable X-ray units, may have high protective conductor leakage currents, and with the increased use of IT equipment, increased total protective conductor currents are possible.

9.19 Isolation and switching: general

710.537.1 Automatic changeover devices need to comply with BS EN 60947-6-1 and be arranged so that safe separation between supply lines is maintained.

9.20 Socket-outlet circuits in the medical IT system for Group 2 medical locations

710.553.1 Socket-outlets intended to supply ME equipment shall be unswitched to prevent the accidental turning-off of vital equipment. The configuration of socket-outlets at each patient treatment position is as follows:

(a) each socket-outlet supplied by an individually protected circuit; or
(b) several socket-outlets separately supplied by a minimum of two circuits.

It is also required that socket-outlets used on medical IT systems shall be coloured blue and be clearly and permanently marked *'Medical Equipment Only'*.

Where two circuits are used, these can be fed from the same IT transformer, where each circuit is individually protected. Additional circuits may be required, due to load considerations, which may also require additional IT transformers.

It is also important that the designer should consider the anticipated use of the socket-outlets and any equipment connected, including the assessment of any loads and their associated inrush currents.

Socket-outlets for TN supplies may be required in Group 2 locations for special equipment, such as mobile X-ray equipment, or for Information Technology equipment located outside the patient environment.

Placing non-medical electrical equipment onto the medical IT system within the patient environment could well compromise safety, due to EMC and leakage current issues (see also Regulations 710.444 and 710.52).

710.559 ## 9.21 Luminaires and lighting installations

In Group 1 and Group 2 medical locations, at least two different sources of supply are required to be provided to maintain safe illumination during medical procedures in case of a fault. One of the sources shall be connected to the electrical supply system for safety services.

9.22 Safety services

710.56 A power supply for safety services is required in order to maintain the supply for continuous operation for a defined period within a pre-set changeover time. The safety power supply system shall automatically take over if the voltage of one or more incoming live conductors, at the main distribution board of the building, has dropped for more than 0.5 s and by more than 10 % in regard to the nominal voltage.

A list of examples with suggested reinstatement times is given in Table 9.1.

9.22.1 Classification of safety services for medical locations

710.560.4 Classification of safety services is given in Regulation 560.4.1 of BS 7671 and shown in Table 9.2.

Safety services provided for locations having differing classifications should meet the classification that gives the highest security of supply.

Refer to Table 9.1 for guidance on the association of classification of safety services with medical locations.

▼ **Table 9.2** Classification of safety services

Classification	Changeover time (s)	Description
No-break[1]	0	Automatic supply available with no break
Very short break	0.15	Automatic supply available within 0.15 s
Short break	0.5	Automatic supply available within 0.5 s
Normal break	5	Automatic supply available within 5 s
Medium break	15	Automatic supply available within 15 s
Long break	> 15	Automatic supply available in more than 15 s

[1]Mains-floating UPS sources satisfy the 'No-break' classification requirement. Other types of UPS source can satisfy the 'Very short break' or 'Short break' classification.

9.22.2 General requirements for safety power supply sources of Group 1 and Group 2 medical locations

710.560.5.5 Primary cells are not allowed as safety power sources.

An additional main incoming power supply, from the general power supply, is not regarded as a source of the safety power supply.

The availability (readiness for service) of safety power sources shall be monitored and indicated at a suitable location.

9.22.3 Failure of the general power supply source

710.560.5.6 In case of a failure of the general power supply source, referred to as the primary electrical supply in HTM 06-01, the power supply for safety services shall be energized to feed the equipment stated in Regulations 710.560.6.1.1, 710.560.6.1.2 and 710.560.6.1.3 with electrical energy for a defined period of time and within a predetermined changeover period, to provide the level of safety needed.

9.22.4 Socket-outlets

710.560.5.7 Where socket-outlets are supplied from the safety power supply source, they shall be readily identifiable according to their safety services classification.

This identification can be by colour coding or/and text.

9.22.5 Detailed requirements for safety power supply services

710.560.6.1 Also refer to Regulation 710.560.5.5.

9.22.6 Power supply sources with a changeover period less than or equal to 0.5 s

710.560.6.1.1 In the event of a voltage failure on one or more line conductors at the distribution board, a safety power supply source shall be used and be capable of providing power for a period of at least 3 h for the following:

(a) luminaires of operating theatre tables;
(b) ME equipment containing light sources being essential for the application of the equipment, such as endoscopes, and including associated essential equipment, such as monitors; and
(c) life-supporting ME equipment.

The duration of 3 h may be reduced to 1 h for items (b) and (c) if a power source meeting the requirements of Regulation 710.560.6.1.2 is installed. The normal power supply should be restored within a changeover period not exceeding 0.5 s.

Supporting information relating to the autonomy of battery inverter units for theatre luminaires is given in HTM 06-01.

9.22.7 Power supply sources with a changeover period less than or equal to 15 s

710.560.6.1.2 Equipment listed under the requirements of safety lighting in Regulation 710.560.9.1 and other services in Regulation 710.560.11 shall be connected within 15 s to a safety power supply source capable of maintaining it for a minimum period of 24 h. This must occur when the voltage of one or more live conductors at the main distribution board for the safety services has decreased by more than 10 % of the nominal value of supply voltage, for a duration greater than 3 s.

Although not explicitly stated, this will include ME equipment requiring a classification of ≤ 15 seconds, as listed in Table 9.1.

9.22.8 Power supply sources with a changeover period greater than 15 s

710.560.6.1.3 Equipment, other than that covered by Regulations 710.560.6.1.1 and 710.560.6.1.2, which is required for the maintenance of healthcare installations, shall be connected either automatically or manually to a safety power supply source capable of maintaining it for a minimum period of 24 h. Examples of this equipment include:

(a) sterilization equipment;
(b) technical building installations, in particular air-conditioning, heating and ventilation systems, building services and waste disposal systems;
(c) cooling equipment;
(d) catering equipment; and
(e) storage battery chargers.

9.22.9 Circuits of safety services

710.560.7 The circuit that connects the power supply source for safety services to the main distribution board is considered a safety circuit.

See Regulation 560.7 of BS 7671 for details of the requirements associated with circuits for safety services.

9.22.10 Safety lighting

710.560.9.1 Lighting is clearly an important safety consideration for medical locations. Regulation 710.560.9.1 requires that in the event of mains power failure, the changeover period to the safety services source shall not exceed 15 s. It also states the necessary minimum illuminance provided for the following:

(a) emergency lighting and exit signs.
(b) locations for switchgear and controlgear for emergency generating sets, for main distribution boards of the normal power supply and for power supply for safety services.
(c) rooms in which essential services are intended. In each such room, at least one luminaire shall be supplied from the power source for safety services.
(d) locations of central fire alarm and monitoring systems.
(e) rooms of Group 1 medical locations: in each such room, at least one luminaire shall be supplied from the power supply source for safety services.
(f) rooms of Group 2 medical locations: a minimum of 90 % of the lighting shall be supplied from the power source for safety services.

The luminaires of the escape routes must be arranged on alternate circuits.

9.22.11 Other services

710.560.11 Other services that may require a safety service supply with a changeover period not exceeding 15 s include, for example, the following:

(a) firefighters' lifts.
(b) ventilation systems for smoke extraction.
(c) paging/communication systems.
(d) ME equipment used in Group 2 medical locations that serves for surgical or other procedures of vital importance. Such equipment will be defined by responsible staff.

(e) electrical equipment of medical gas supply, including compressed air, vacuum supply and narcosis (anaesthetics) exhaustion, as well as their monitoring devices.

(f) fire detection and fire alarms.

(g) fire-extinguishing systems.

9.23 Inspection and testing

710.6 The testing of equipment connected to the electrical installation is outside the scope of BS 7671:2018. For ME equipment and ME systems, refer to BS EN 62353.

9.23.1 Initial verification: general

710.641 The dates and results of each verification are required to be recorded.

The tests specified below under items (a) to (c), in addition to the requirements of Chapter 64 of BS 7671:2018, shall be carried out both prior to commissioning and after alteration or repairs and before re-commissioning:

(a) complete functional tests of the IMDs associated with the medical IT system, including insulation failure, transformer high temperature, overload, discontinuity and the audible and/or visual alarms linked to them;

(b) measurements of leakage current of the output circuit and of the enclosure of the medical IT transformers in no-load condition, as specified in Regulation 710.512.1.1; and

(c) measurements to verify that the resistance of the supplementary equipotential bonding is within the limits specified in Regulation 710.415.2.2.

The tests specified within HTM 06-01 may also be required as part of client requirements.

9.23.2 Periodic inspection and testing: general

710.651 As a guide, in addition to the requirements of Chapter 65 of BS 7671:2018, the following procedures are recommended at the given intervals:

(a) annually: complete functional tests of the IMDs associated with the medical IT system, including insulation failure, transformer high temperature, overload, discontinuity and the audible/visual alarms linked to them;

(b) annually: measurements to verify that the resistance of the supplementary equipotential bonding is within the limits specified in Regulation 710.415.2.2; and

(c) every three years: measurements of leakage current of the output circuit and of the enclosure of the medical IT transformers in no-load condition, as specified in Regulation 710.512.1.1.

Client or local Health Authority requirements, if any, may apply. Supporting information on periodic inspection and testing is also given in HTM 06-01.

9.24 Further guidance

9.24.1 The Department of Health and Social Care

The Department of Health and Social Care provides medical, building and engineering guidance through its Health Technical Memorandum (HTM) series of documents. In Scotland and Wales, these documents are referred to as Scottish Health Technical Memoranda (SHTM) and Welsh Health Technical Memoranda (WHTM), respectively. The table below lists some items of interest obtained from:

https://www.gov.uk/government/publications/guidance-on-electrical-services-supply-and-distribution-within-healthcare-premises

HTM 06-01 (April 2017)	Electrical Services supply and distribution:
	Provides guidance for all works on the fixed wiring and integral electrical equipment used for electrical services within healthcare premises. The document should be used for all forms of electrical design work ranging from a new "greenfield" site to modifying an existing installation.
HTM 06-02	Electrical Safety Guidance for low voltage systems (2006)
HTM 06-02	Electrical Safety Handbook
HTM 08-03	Management of Bedhead services in the health sector

9.24.2 IET publications

Useful information and guidance can be found in the IET publication *Guide to Electrical Installations in Medical Locations*.

Solar photovoltaic (PV) power supply systems

10

10.1 Introduction

Sect 712 More and more solar photovoltaic (PV) systems are being installed with the aim of cutting carbon dioxide emissions and reducing electricity bills. Section 712 of BS 7671 has specific requirements for the installation of PV systems.

Note: A new IEC standard, IEC 60364-7-712:2017, which applies to the electrical installation of PV systems intended to supply all or part of an installation, has been published. In addition, HD 60364-7-712, dated April 2016, has been published.

An IET Code of Practice on photovoltaic systems, *Code of Practice for Grid Connected Solar Photovoltaic (PV) systems*, published in March 2015, is available. In addition, a guide to the installation of photovoltaic systems, published by the Microgeneration Certification Scheme, is available.

There are no significant changes to Section 712 introduced by the 18th Edition.

10.2 Scope

712.1 The particular requirements of Section 712 apply to the electrical installations of PV power supply systems including systems with AC modules.

Note: Requirements for PV power supply systems that are intended for stand-alone operation are under consideration.

10.3 The Electricity Safety, Quality and Continuity Regulations 2002

The solar photovoltaic (PV) power supply systems described in this chapter are required to meet the requirements of the Electricity Safety, Quality and Continuity Regulations 2002 (ESQCR), as they are embedded generators.

Where the output does not exceed 16 A per phase they are, by definition, small-scale embedded generators (SSEG) and are exempted from certain of the requirements of Regulation 22 provided that:

(a) the equipment is type tested and approved by a recognized body;
(b) the consumer's installation complies with the requirements of BS 7671;
(c) the equipment will disconnect itself from the distributor's network in the event of a network fault; and
(d) the distributor is advised of the installation before or at the time of commissioning.

Installations will need to meet the requirements of the Energy Networks Association Engineering Recommendation G83, *Recommendations for the connection of small-scale embedded generators (up to 16 A per phase) in parallel with the public low voltage distribution network.*

Note: A new Engineering Recommendation G98 was published on the 16 May 2018.

G98 supersedes Engineering Recommendation G83 from 27 April 2019. Until then, persons can use either G83 or G98. On 27 April 2019 G83 will be withdrawn and all new generation from that date must comply with the requirements of G98.

Engineering Recommendation G98 contains requirements for the connection of Fully Type Tested Micro-generators (up to and including 16 A per phase) in parallel with public Low Voltage Distribution Networks on or after the 27 April 2019.

The requirements of the ESQCR for small-scale embedded generators are discussed more fully in Chapter 15.

For generator installations larger than 16 A per phase see Energy Networks Association Engineering Recommendation G59.

Note: A new Engineering Recommendation G99 has been published on 16 May 2018.

G99 supersedes Engineering Recommendation G59 from 27 April 2019. Until then persons can use either G59 or G99. On 27 April 2019 G59 will be withdrawn and all new generation from that date must comply with the requirements of G99.

10.4 Normative references

The following equipment standard for PV modules is referred to in Section 712 of BS 761:

▶ IEC 61215: *Crystalline PV modules – Design qualification and type approval*

Note: This is not an exhaustive list.

10.5 Definitions

Part 2 (See also Figures 10.1 and 10.2.)

PV AC module: integrated module/convertor assembly where the electrical interface terminals are AC only. No access is provided to the DC side.

PV array: mechanically and electrically integrated assembly of PV modules, and other necessary components, to form a DC power supply unit.

PV array cable: output cable of a PV array.

PV array junction box: enclosure where PV strings of any PV array are electrically connected and where devices can be located.

PV cell: basic PV device which can generate electricity when exposed to light such as solar radiation.

PV convertor: device which converts DC voltage and DC current into AC voltage and AC current.

PV DC main cable: cable connecting the PV generator junction box to the DC terminals of the PV convertor.

PV generator: assembly of PV arrays.

PV generator junction box: enclosure where PV arrays are electrically connected and where devices can be located.

PV installation: erected equipment of a PV power supply system.

PV module: smallest completely environmentally protected assembly of interconnected PV cells.

PV string: circuit in which PV modules are connected in series, in order for a PV array to generate the required output voltage.

PV string cable: cable connecting PV modules to form a PV string.

PV supply cable: cable connecting the AC terminals of the PV convertor to a distribution circuit of the electrical installation.

Standard test conditions (STC): test conditions specified in BS EN 60904-3 for PV cells and PV modules.

Open-circuit voltage under standard test conditions U_{OC} STC: voltage under standard test conditions across an unloaded (open) generator or on the DC side of the convertor.

Short-circuit current under standard test conditions I_{SC} STC: short-circuit current of a PV module, PV string, PV array or PV generator under standard test conditions.

DC side (not a BS 7671 definition): part of a PV installation from a PV cell to the DC terminals of the PV convertor.

AC side (not a BS 7671 definition): part of a PV installation from the AC terminals of the PV convertor to the point of connection of the PV supply cable to the electrical installation.

Simple separation: separation between circuits or between a circuit and Earth by means of basic insulation.

10.6 Protection for safety

10.6.1 Protection against electric shock

712.410.3   PV equipment on the DC side is to be considered energized, even when the system is disconnected from the AC side.

Protection by extra-low voltage: SELV and PELV

712.414.1.1   For SELV and PELV systems, U_{OC} STC replaces U_0 and shall not exceed 120 V DC.

10.6.2 Fault protection

712.411 **(a) Protection by automatic disconnection of supply**

Note: Protection by automatic disconnection of supply on the DC side requires special measures that are under consideration at European level.

712.411.3.2.1.1 On the AC side, the PV supply cable shall be connected to the supply side of the protective device for automatic disconnection of circuits supplying current-using equipment.

712.411.3.2.1.2 Where an electrical installation includes a PV power supply system without at least simple separation between the AC side and the DC side, an RCD installed to provide either fault protection by automatic disconnection of supply or additional protection in accordance with Regulation 415.1.1, for the PV supply cable shall be type B to BS EN 62423. Where the PV power supply convertor by construction is not able to feed DC fault currents into the electrical installation, an RCD of type B to BS EN 62423 is not required. (A type B RCD ensures tripping for residual sinusoidal alternating currents, pulsating direct currents and smooth direct currents, whether suddenly applied or slowly rising. For special applications, refer to the manufacturer.)

The flow diagram below shows a typical decision-making process that could be used:

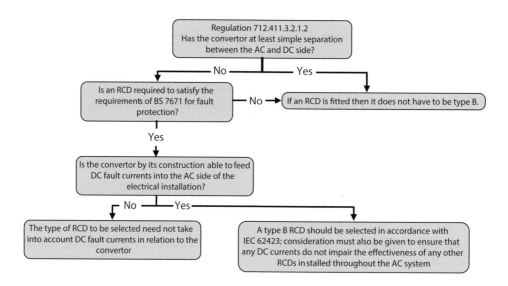

(b) PV connection to distribution boards/consumer units containing RCDs
It is recognized that where a residual current circuit-breaker (RCCB) is double-pole, then for a line to earth or neutral to earth fault, automatic disconnection of the fault current would be achieved. It is noted that the PV supply would be maintained to the loads for up to five seconds; however, the fault current would be disconnected.

The following points should be noted:

(i) double-pole RCCBs that are not marked line or load, or by arrows indicating the direction of power flow, can be connected supply or load side from a PV supply.

(ii) double-pole RCCBs that are marked line or load, or by arrows indicating the direction of power flow, can only be connected supply side from a PV supply.

(iii) if all final circuits are protected by residual current circuit-breakers (with overcurrent protection) (RCBO), they can be single-pole. The direction of power flow will be correct, with a dedicated PV circuit supply to the distribution busbar.

(iv) if an RCBO is used to connect a PV supply to the distribution busbar, its direction of p   ower flow (line and load marking) must be observed.

(c) Other protective measures

712.412 Protection by use of Class II or equivalent insulation should preferably be adopted on the DC side.

712.410.3.6 Protection by non-conducting location (Regulation 418.1) and earth-free local equipotential bonding (Regulation 418.2) shall not be used on the DC side.

10.6.3 Protection against overload on the DC side

Note: For PV cables on the DC side not complying with the paragraphs below, the requirements of Chapter 43 of BS 7671 apply for overload protection.

712.433.1 Overload protection may be omitted to PV string and PV array cables when the continuous current-carrying capacity of the cable is equal to or greater than 1.25 times I_{sc} STC at any location.

712.433.2 Overload protection may be omitted to the PV main cable when the continuous current-carrying capacity of the PV main cable is equal to or greater than 1.25 times I_{sc} STC of the PV generator.

Note: The above requirements are only relevant for the protection of the cables. It is important to consult the manufacturer for the protection of the modules. To protect against the risk of module fires overload protection needs to be installed. The selection of overcurrent protective measures depends upon the system design and the number of strings. For advice on fuse selection refer to the IET Code of Practice on photovoltaic systems, *Code of Practice for Grid Connected Solar Photovoltaic (PV) Systems*, or the guide to the installation of Photovoltaic systems published by the Microgeneration Certification Scheme. Also refer to IEC 60269-6 *Low-voltage fuses – Part 6: Supplementary requirements for fuse-links for the protection of solar photovoltaic energy systems*.

Refer to the manufacturer for more information.

10.6.4 Protection against fault current

712.434.1 The PV supply cable on the AC side shall be protected against fault current by an overcurrent protective device installed at the connection to the AC mains.

10.6.5 Protection against electromagnetic interference (EMI)

712.444.4.4 To minimize voltages induced by lightning, the area of all wiring loops must be kept as small as possible.

10.7 Isolation and switching

712.537.2.1.1 To allow maintenance of the PV convertor, means of isolating the convertor from the DC side and the AC side shall be provided.

> **Note:** Further requirements about the isolation of a PV installation operating in parallel with the public supply system are given in Regulation 551.7.6 of BS 7671; see section 15.4 in Chapter 15.

10.8 Selection and erection of equipment

10.8.1 Compliance with standards

712.511.1 PV modules shall comply with the requirements of the relevant equipment standard, e.g. BS EN 61215 for crystalline PV modules. PV modules of Class II construction or with equivalent insulation are recommended if the U_{OC} STC of the PV strings exceeds 120 V DC.

The PV array junction box, PV generator junction box and switchgear assemblies shall be in compliance with BS EN 61439-1.

The designer and installer should verify that the cables are suitable for the installation conditions. Cable designed and tested specifically for use in PV systems is available.

It is understood that an IEC standard for PV string cables is under development. It is recommended that in the interim cables should comply with UL 4703 or TUV 2 Pfg 1169 08.2007.

10.8.2 Operational conditions and external influences

712.512 Electrical equipment on the DC side shall be suitable for direct voltage and direct current.

PV modules may be connected in series up to the maximum allowed operating voltage of the PV modules and the PV convertor, whichever is lower. Specifications for this equipment shall be obtained from the equipment manufacturer.

If blocking diodes are used, their reverse voltage shall be rated for $2 \times U_{OC}$ STC of the PV string. The blocking diodes shall be connected in series with the PV strings (see Figures 10.1 and 10.2).

712.512.2.1 When installing PV modules the installer must follow the manufacturer's instructions for mounting so that adequate heat dissipation is provided under the conditions of maximum solar radiation to be expected. Such instructions are required by the equipment standard.

10.8.3 Accessibility

712.513.1 The selection and erection of equipment shall facilitate safe maintenance and shall not adversely affect provisions made by the manufacturer of the PV equipment to enable maintenance or service work to be carried out safely.

10.8.4 Wiring systems

712.522 Selection and erection in relation to external influences

PV string cables, PV array cables and PV DC main cables shall be selected and erected so as to minimize the risk of earth faults and short-circuits.

Note: This may be achieved by, for example, reinforcing the protection of the wiring against external influences by the use of single-core sheathed cables, complying with BS EN 50618.

Wiring systems shall withstand the expected external influences such as wind, ice formation, temperature and solar radiation.

10.8.5 Devices for isolation and switching

712.537.2.2.1
514.15.1 In the selection and erection of devices for isolation and switching to be installed between the PV installation and the public supply, the public supply shall be considered the source and the PV installation shall be considered the load. A notice that warns of the two sources of supply should be provided at each point of isolation.

712.537.2.2.5 A switch disconnector shall be provided on the DC side of the PV convertor.

All junction boxes (PV generator and PV array boxes) shall carry a warning label indicating that active parts inside the boxes may still be live after isolation from the PV convertor.

WARNING
PV SYSTEM
Parts inside this box or enclosure may still be live after isolation from the supply.

10.9 Earthing arrangements and protective conductors

712.54 Where protective bonding conductors are installed, they shall be parallel to and in as close contact as possible with DC cables and AC cables and accessories.

The flow diagram below shows a typical decision-making process that could be used:

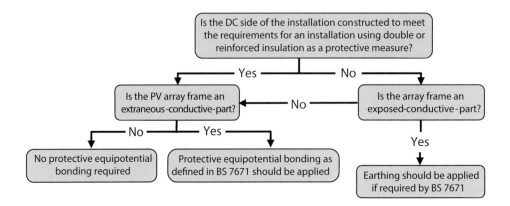

Note: Functional earthing of the array frame may be required to ensure the PV system functions correctly. The designer should seek advice from the manufacturer on the functional earthing requirements. Also, see the IET *Code of Practice for Grid Connected Solar Photovoltaic Systems.*

10.9.1 Types of system earthing

712.312.2 Engineering Recommendation G83 published by the Energy Networks Association makes it clear that when an SSEG is operating in parallel with the distribution system of a distribution network operator (DNO), there shall be no direct connection between the SSEG current-carrying conductors and earth, with the following exception: G83 allows the connection of one pole of the DC side of the convertor to the DNO's earth terminal, subject to certain conditions. For more information, refer to the manufacturer.

Note: A new Engineering Recommendation G98 has now been published. G98 supersedes Engineering Recommendation G83. Engineering Recommendation G98 contains requirements for the connection of Fully Type Tested Micro-generators (up to and including 16 A per phase) in parallel with public Low Voltage Distribution Networks on or after the 17 May 2019.

Regulation 712.312.2 allows earthing of one of the live conductors of the DC side, if there is at least simple separation between the AC side and the DC side.

Note: Any connections with Earth on the DC side should be electrically connected, to avoid corrosion.

A risk assessment should be carried out to determine exposed-conductive-parts and extraneous-conductive-parts.

▶ **Figure 10.1** PV installation – general schematic, one array

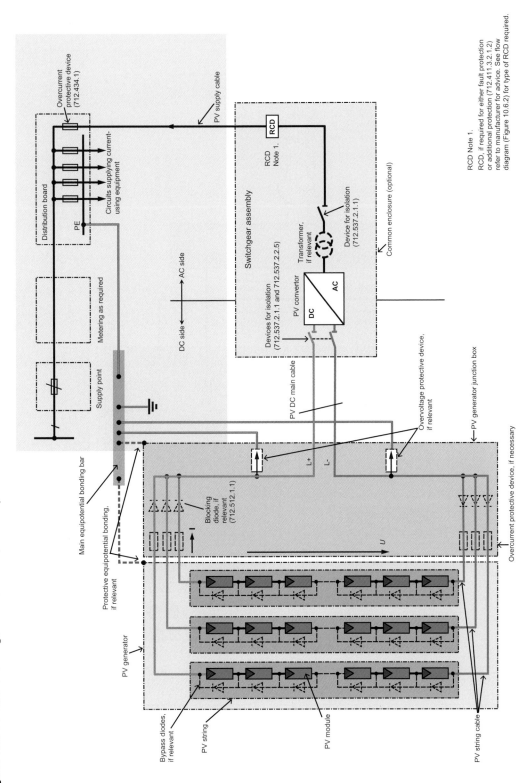

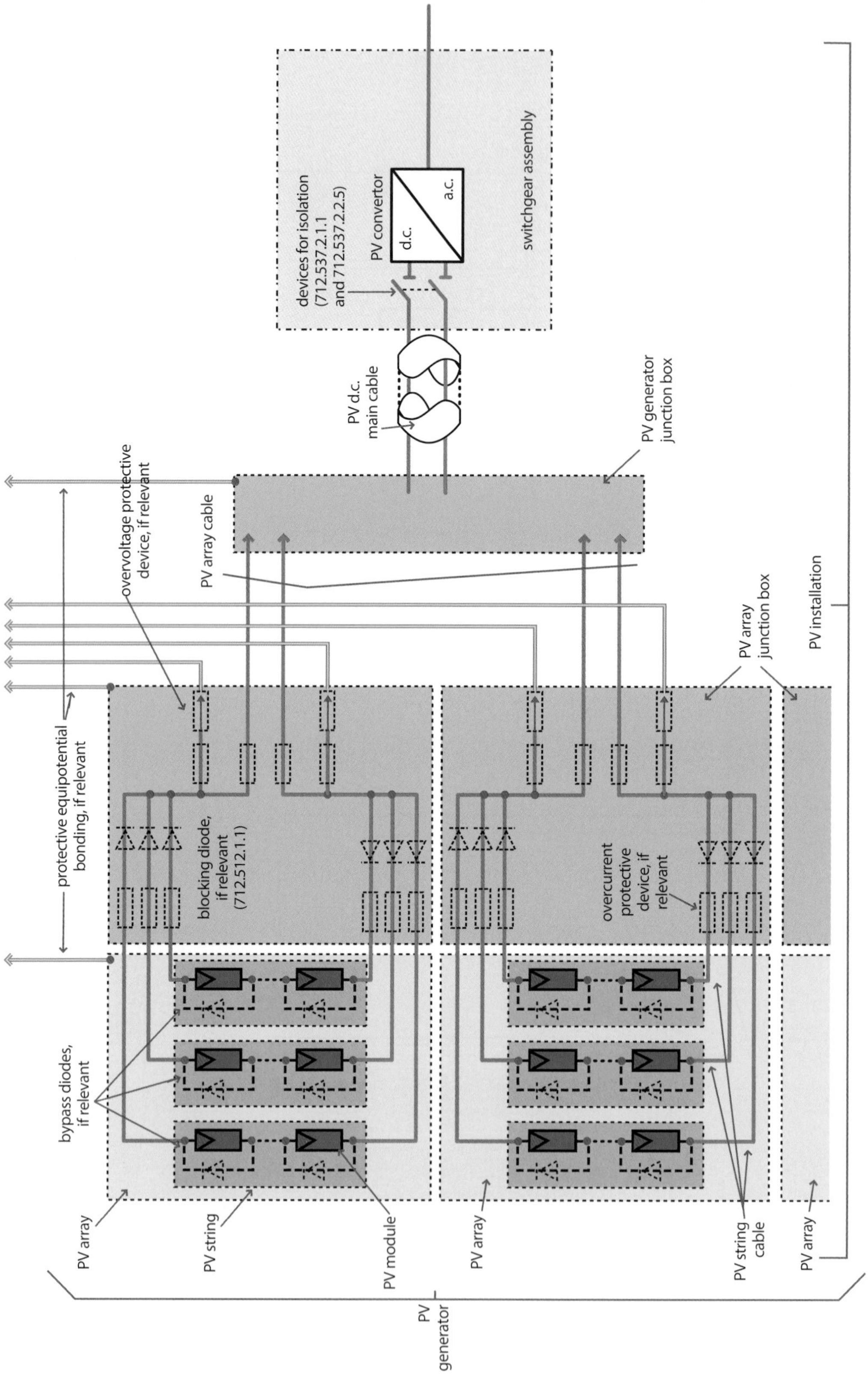

▶ **Figure 10.2** PV installation – example with two or more arrays

Exhibitions, shows and stands

11

11.1 Introduction

Sect 711 The 18th Edition introduces only minor changes to Section 711. Regulation 711.55.1.5 has been deleted, as the requirements are covered by the general rules in Chapter 42 (see Section 11.5).

11.2 Scope

Part 2 Section 711 of BS 7671:2018 is concerned with temporary electrical installations in exhibitions, shows and stands (including mobile and portable displays and equipment). Such installations may be installed indoors or outdoors within permanent or temporary structures. It does not apply to the fixed electrical installation of the building, if any, in which the exhibition, show or stand takes place.

Section 711 does not apply to electrical systems as defined in BS 7909 used in structures, sets, mobile units etc. as used for public or private events, touring shows, theatrical, radio, TV or film productions and similar activities of the entertainment industry.

11.3 The risks

The particular risks associated with exhibitions, shows and stands are those of electric shock and fire. These arise from:

(a) the temporary nature of the installation;
(b) the lack of permanent structures;
(c) the severe mechanical stresses; and
(d) access to the general public.

Because of these particular risks, additional measures are recommended as described in the following sections.

11.4 Protection against electric shock

711.410.3.4 A cable intended to supply temporary structures shall be protected at its origin by an RCD whose rated residual operating current does not exceed 300 mA. This device shall provide a delay by using a device in accordance with BS EN 60947-2, or be of the S-type in accordance with BS EN 61008-1 or BS EN 61009-1 for selectivity with RCDs protecting final circuits. See Figure 11.1.

> **Note:** The requirement for additional protection relates to the increased risk of damage to cables in temporary locations.

711.410.3.5 Protective measures against electric shock by means of obstacles and by placing out of reach (Section 417) shall not be used.

711.410.3.6 The protective measures of non-conducting location (Regulation 418.1) and earth-free local equipotential bonding (Regulation 418.2) shall not be used.

11.4.1 Protection by automatic disconnection of supply

711.411.4 **TN system**

Except for a part of an installation within a building, a PME earthing facility shall not be used as the means of earthing for an installation falling within the scope of Section 711 except where:

(a) the installation is continuously under the supervision of a skilled or instructed person(s); and

(b) the suitability and effectiveness of the means of earthing has been confirmed before the connection is made.

Where these conditions cannot be met, if the exhibition supply is derived from a building that is supplied by a PME system then the exhibition supply will have to be part of a TT system having a separate connection to Earth independent from the PME earthing.

11.4.2 Additional protection

711.411.3.3 Each socket-outlet circuit not exceeding 32 A and all final circuits other than for emergency lighting shall be protected by an RCD having the characteristics specified in Regulation 415.1.1. However, consideration has to be given to the hazards of loss of lighting in such a public place, particularly when crowded. Lighting of such areas should always be on at least two separate circuits with separate RCDs, and should preferably be out of reach of the general public (see Figure 11.1).

Note: A new Engineering Recommendation G98 was published on 16 May 2018.

G98 supersedes Engineering Recommendation G83 from 27 April 2019. Until then persons can use either G83 or G98. On 27 April 2019 G83 will be withdrawn and all new generation from that date must comply with the requirements of G98.

Engineering Recommendation G98 contains requirements for the connection of Fully Type Tested Micro-generators (up to and including 16 A per phase) in parallel with public Low Voltage Distribution Networks on or after 27 April 2019.

In addition a new Engineering Recommendation G99 has been published on 16 May 2018.

G99 supersedes Engineering Recommendation G59 from 27 April 2019. Until then persons can use either G59 or G99. On 27 April 2019 G59 will be withdrawn and all new generation from that date must comply with the requirements of G99.

▼ **Figure 11.1** Exhibition/show distribution with standby generator

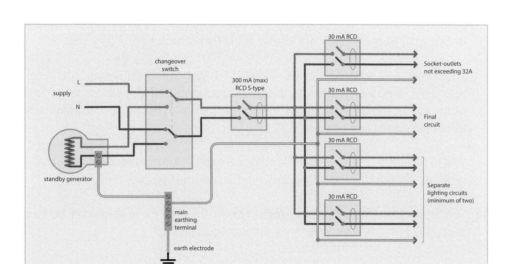

Note 1: Regulation 21 of the ESQCR has requirements for switched alternative sources of energy; see section 11.9.

Note 2: For clarity, generator controls, overcurrent protection and isolation not fully shown.

11.4.3 Bonding of stands, vehicles, wagons, caravans and containers

711.411.3.1.2 Structural metallic parts that are accessible from within the stand, vehicle, wagon, caravan or container shall be connected through the main protective bonding conductors to the main earthing terminal within the unit.

11.4.4 Protection by SELV or PELV

711.414.4.5 Where SELV or PELV installations are used, whatever the nominal voltage, basic protection is required to be provided by basic insulation, or by barriers or enclosures providing protection of at least IPXXD or IP4X rating.

11.5 Protection against thermal effects

Chap 42 There is often an increased risk of fire and burns in temporary locations. For this reason it is important to comply with all the relevant requirements of Chapter 42.

11.5.1 Protection against fire

Installation designers must bear in mind that stored materials may present a particular hazard in such a location, particularly fodder, cardboard boxes etc.

422.3.7 Motors that are automatically or remotely controlled and are not continuously supervised must be fitted with manual reset devices for protection against excess temperature.

11.5.2 Lighting

711.422.4.2 Lighting equipment such as incandescent lamps, spotlights and small projectors, and
422.3.1 other equipment or appliances with high temperature surfaces, should, in addition
559.4.1 to being suitably guarded, be arranged well away from combustible materials such as to prevent contact. Equipment should be installed and located in accordance with relevant standards and manufacturers' instructions.

711.559.5 Luminaires that are mounted below a height of 2.5 m from floor or ground level, or mounted in a position that is accessible to accidental contact, should be firmly and adequately fixed, and sited or guarded so as to prevent the risk of injury to persons or ignition of materials. Where this includes outdoor lighting installations the requirements of Section 714 of BS 7671 would also apply (see Chapter 18).

Showcases and signs should be constructed from materials having adequate heat resistance, mechanical strength, electrical insulation and ventilation, taking account of the combustibility of exhibits in relation to the heat generated. The manufacturer's instructions must be complied with.

Where there is a concentration of electrical equipment, including luminaires, that might generate considerable heat, adequate ventilation must be provided.

GN4 Similar guidance for protection against fire for this type of location can be found in section 8.3 of Guidance Note 4: *Protection Against Fire*.

11.6 Isolation

711.537.2.3 Every separate temporary structure, such as a vehicle, stand or unit, intended to
Sect 462 be occupied by one specific user, and each distribution circuit supplying outdoor
537.2 installations, should be provided with its own readily accessible and properly identifiable means of isolation. Switches, circuit-breakers and RCDs etc. considered suitable for isolation by the relevant standard or the manufacturer may be used.

11.7 Measures of protection against overcurrent

430.3 All circuits must be protected against overcurrent by a suitable protective device located at the origin of the circuit.

11.8 Selection and erection of equipment

711.51 Control and protective switchgear must be placed in closed cabinets that can only be
Part 2 opened by a key or tool, except those parts that are designed and intended to be operated by ordinary persons.

512.2 Equipment, particularly switchgear and fusegear, must be mounted away from locations that may not be weatherproof. Tent poles etc. where used for mounting switchgear are often the weak point in the weather tightness of temporary structures.

The means of isolation for each stand or unit should not be locked away and should be readily accessible and obvious to the stand user (see section 11.6).

11.8.1 Wiring systems

Sect 522 Particular care must be paid to the selection and installation of cables to ensure that
Sect 523 the mechanical protection, insulation, heat resistance and current-carrying capacity are sufficient for the conditions likely to be encountered.

711.52 Wiring cables shall be copper, have a minimum cross-sectional area of 1.5 mm^2, and shall comply with an appropriate British or Harmonized Standard for either thermoplastic or thermosetting insulated electric cables. Flexible cables shall not be laid in areas accessible to the public unless they are protected against mechanical damage, and do not present a trip hazard.

Mechanical protection or armoured cables should be used wherever there is a risk of mechanical damage.

Underground cables are susceptible to damage by structure support pins that may be up to 1 m in length. Exhibitors must be advised of the presence of cables and, if necessary, the cable route marked. The general rules for buried cables must be followed; see also section 13.3 in Chapter 13.

711.521 Where a building is used for exhibitions etc. and does not include a fire alarm system, only the following cable systems should be used:

(a) flame retardant cables to BS EN 60332-1-2 or BS EN 60332-3, and low smoke to BS EN 61034-2;

(b) single-core or multicore cables enclosed in metallic or non-metallic conduit or trunking that provides fire protection in accordance with the BS EN 61386 series or the BS EN 50085 series. This is to provide a degree of protection of at least IP4X.

11.8.2 Electrical connections
711.526.1 Joints must not be made in cables, unless as a connection into a circuit.

Conductor connections should be within an enclosure providing a degree of protection of at least IPXXD or IP4X rating and the enclosure should incorporate a suitable cable anchorage.

11.8.3 Lighting installations
ELV lighting systems for filament lamps
711.559.4.2 ELV systems for filament lamps must comply with BS EN 60598-2-23.

Lampholders
711.559.4.3 Insulation piercing lampholders should not be used unless the cables and lampholders are compatible and the lampholders are non-removable once fitted to the cable.

Electric discharge lamp installations
711.559.4.4 Installations of any luminous tube, sign or lamp as an illuminated unit on a stand, or as an exhibit, with nominal power supply voltage higher than 230/400 V AC, must comply with Regulations 711.559.4.4.1 to 711.559.4.4.3.

The sign or lamp should be installed out of arm's reach or be adequately protected to reduce the risk of injury to persons.

The facia or stand fitting material behind luminous tubes, signs or lamps must be non-ignitable.

Emergency switching device
711.559.4.4.3 A separate circuit should be used to supply signs, lamps or exhibits and should be controlled by an emergency switch. The switch must be easily visible, accessible and clearly marked.

Protection against thermal effects
711.559.5 Luminaires mounted below 2.5 m (arm's reach) from floor level or otherwise accessible to accidental contact must be firmly and adequately fixed, and sited or guarded so as to prevent risk of injury to persons or ignition of materials.

11.8.4 Electric motors

711.55.4.1
465.1
Where an electric motor might give rise to a hazard the motor must be provided with an effective means of isolation on all poles and where necessary an emergency stop, and the means should be adjacent to the motor it controls (see BS EN 60204-1).

11.8.5 ELV transformers and electronic convertors

711.55.6
A manual reset protective device must protect the secondary circuit of each transformer or electronic convertor.

Particular care must be taken when installing ELV transformers, which shall be mounted out of arm's reach of the public, e.g. in a panel or room with adequate ventilation that can only be accessed by skilled or instructed persons. Such access shall be provided only to facilitate inspection, testing and maintenance.

Electronic convertors must conform with BS EN 61347-1.

11.8.6 Socket-outlets

711.55.7
Floor mounted socket-outlets should preferably not be installed. Where their use is unavoidable, they must be adequately protected from mechanical damage and ingress of water.

11.9 Generators

Sect 551
Installations incorporating generator sets must comply with Section 551 of BS 7671. Where a generator is used to supply the temporary installation using a TN or TT system, it must be ensured that the installation is earthed, preferably by separate earth electrodes. For TN systems all exposed-conductive-parts should be bonded back to the generator. The neutral conductor and/or star point of the generator should be connected to the exposed-conductive-parts of the generator and reference earthed.

Part VI of the ESQCR provides requirements for generation. Regulation 21 has requirements for switched alternative sources of energy (see Figure 11.1) as follows:

Where a person operates a source of energy as a switched alternative to a distributor's network, he shall ensure that that source of energy cannot operate in parallel with that network and where the source of energy is part of a low voltage consumer's installation, that installation shall comply with British Standard Requirements [meaning BS 7671].

The requirements for parallel operation are much more onerous.

11.10 Safety services

Chapter 56
Where an exhibition is held within a building, it is assumed that the emergency lighting and/or fire safety systems etc. will be part of the permanent installation within that building. Care should be taken to ensure that existing emergency escape signs and escape routes are not obscured, impeded or blocked.

Additional emergency lighting should be installed in those areas not covered by the permanent installation. Where an exhibition is constructed out of doors, an adequate fire alarm system should be installed to enclosed areas to facilitate emergency evacuation.

Where an event is taking place out of doors and is open to the public in partial or total darkness, then:

(a) emergency lighting should be provided to escape routes; and

(b) provision should be made to ensure that alternative sources of supply for general lighting sufficient for safe evacuation are available throughout the area.

Chapter 56
Appx 1

Reference should be made to the following British Standards:

▶ BS 5266 series and BS EN 1838 *Emergency lighting*; and

▶ BS 5839 series *Fire detection and fire alarm systems for buildings*.

11.11 Inspection and testing

711.6 All temporary electrical installations should be retested on site in accordance with Chapter 64 of BS 7671, after each assembly on site.

Users (e.g. exhibitors and stall holders) should be advised to visually check electrical equipment for damage on a daily basis.

Guidance Note 7: Special Locations
© The Institution of Engineering and Technology

Heating cables and embedded heating systems

12

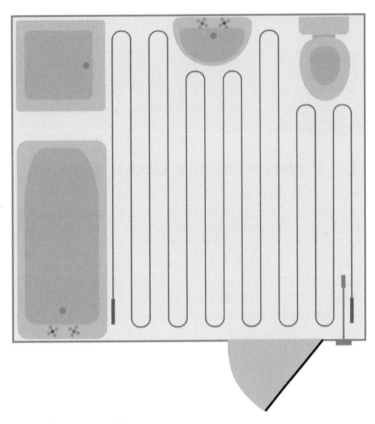

[Illustration courtesy of Warmup plc]

12.1 Introduction

Sect 753 The scope of Section 753 has been extended for the 18th Edition, based on a new international standard: IEC 60364-7-753.

12.2 Scope

753.1 The scope of Section 753 now applies to embedded electric heating systems for surface heating. It also applies to electric heating systems for de-icing or frost prevention or similar applications, and covers both indoor and outdoor systems. These include heating systems for walls, ceilings, floors, roofs, drainpipes, gutters, pipes, stairs, roadways and non-hardened compacted areas (for example, football fields and lawns).

Heating systems for industrial and commercial applications complying with BS EN 60519, BS EN 62395 and BS EN 60079 are not covered.

Consequently, Section 753 now includes additional requirements to cover wall heating, heating conductors and cables laid in soil and concrete, etc. Additional requirements are included to cover the prevention of mutual detrimental influence. Documentation is also covered. The designer will be required to provide appropriate information about approved substances in the surroundings of the heating units.

12.3 The risks

753.415.1
753.424.101
Among the risks associated with heating systems are penetration of the heating elements by nails, screws and the like. For this reason, additional protection is required by the use of a 30 mA RCD having the characteristic specified in Regulation 415.1.1. Time-delayed type RCDs shall not be used.

There is a risk of fire to the building structure due to the use of heating. Consequently, there are requirements to prevent the heating elements creating high temperatures in adjacent material.

12.4 Protection against electric shock

753.410.3
753.413
753.4113.2
The protective measures of obstacles, placing out of reach, non-conducting location and earth-free local equipotential bonding shall not be used. Also, the protective measure of electrical separation shall not be used for wall heating systems.

Under the protective measure of automatic disconnection of supply, Regulation 753.411.3.2 states that RCDs having the characteristics specified in Regulation 415.1.1 shall be used as disconnecting devices. The regulation requires that in the case of heating units which are delivered from the manufacturer without exposed-conductive-parts, a suitable conductive covering, for example, a metal grid with a spacing of not more than 30 mm, must be provided on site as an exposed-conductive-part above the floor heating elements or under the ceiling heating elements, and connected to the protective conductor of the electrical installation.

Note to Regulation 753.411.3.2 states that where Class II floor or ceiling heating units complying with the requirements of Regulations 412.1.1 and 412.2.1.1 are used, the mesh metal grid is not required.

A second note to the regulation advises that limiting the rated heating power to 7.5 kW/230 V or 13 kW/400 V downstream of a 30 mA RCD may avoid unwanted tripping due to leakage current.

753.412
Under the protective measure of double or reinforced insulation, Regulation 753.412.1.201 makes it clear that the mesh metal grid detailed in Regulation 753.411.3.2 is not required. The regulation requires that where this protective measure is used as the sole protective measure for floor or ceiling heating units, complying with the requirements of Regulation 412.2.1.1, the heating-free areas must be readily identifiable. Furthermore, Regulation 753.412.1.201 does not permit the use of this protective measure as the sole protective measure for a wall heating system.

753.415.1   Regulation 753.415.1 requires that circuits supplying heating units must have additional protection by the use of RCDs having the characteristics specified in Regulation 415.1.1. Time-delayed type RCDs must not be used.

753.42   12.5   Protection against thermal effects

12.5.1 Protection against burns (maximum surface temperature of floor)

753.423   In floor areas where contact with skin or footwear is possible, the surface temperature of the floor shall be limited, for example, to no more than 35 °C for floor heating systems. For additional information, reference may be made to CENELEC Guide 29.

12.5.2 Protection against overheating

753.424   Regulation 753.424.101 requires special care to be taken to prevent the heating elements creating high temperatures to adjacent material. This may be achieved by using heating units with temperature self-limiting functions or by separation with heat-resistant materials. The latter may be accomplished by placing the heating units on a metal sheet, in metal conduit or at a distance of at least 10 mm in air from the ignitable structure.

> **Note:**   Dependent on adjacent material it may be prudent to consider a larger separation distance.

Regulation 753.424.201 requires that for floor or ceiling heating systems in buildings, one or more of the following measures shall be applied within the zone where the heating units are installed to limit the temperature and the heating zone to a maximum of 80 °C:

(a) appropriate design of the heating system;
(b) appropriate installation of the heating system; or
(c) use of protective devices.

Heating units shall be connected to the electrical installation via cold tails or suitable terminals.

Heating units must be permanently connected to cold tails, for example, by a crimped connection.

For wall heating systems (which may be more vulnerable than floors and ceilings to penetration), Section 753 contains additional requirements to protect against the effects of overheating caused by a short-circuit between live conductors due to penetration of an embedded heating unit.

753.424.102   Regulation 753.424.102 requires the heating units for wall heating systems to be provided with a metal sheath, metal enclosure or fine mesh metallic grid. The sheath, enclosure or grid shall be connected to the protective conductor of the supply circuit.

12.6   Standards

753.511   Flexible sheet heating elements should comply with BS EN 60335-2-96 and heating cables should comply with IEC 60800.

12.7 Identification

753.514 The designer of the installation/heating system or installer shall provide documentation for each heating system, containing the following details:

(a) manufacturer and type of heating units;
(b) number of heating units installed;
(c) length/area of heating units;
(d) rated power;
(e) surface power density;
(f) layout of the heating units in the form of a sketch, drawing or picture;
(g) position/depth of heating units;
(h) position of junction boxes;
(i) cables, earthed conductive shields and the like;
(j) rated voltage;
(k) rated resistance (cold) of the heating units;
(l) rated current of overcurrent protective device;
(m) rated residual operating current of the RCD;
(n) the insulation resistance of the heating installation and the test voltage used; and
(o) product information, containing provisions about approved materials in contact with the heating units, with necessary instructions for installation.

The documentation shall be fixed to, or adjacent to, the distribution board of the heating system.

Section 12.11 describes advice to be provided by the installer for the user of the installation.

12.8 Heating-free areas

753.520.4 It may be necessary to provide areas of floor, wall or ceiling that are unheated, e.g. where fixtures to the surface would prevent the proper emission of heat.

753.522.4.3 Where heating units are installed, there shall be heating-free areas where drilling and fixing by screws and the like may be carried out without risk of damage to the units.

12.9 Locations containing a bath or shower

701.753 For electric floor heating systems, Regulation 701.753 requires that only heating cables according to relevant product standards or thin sheet flexible heating elements according to the relevant equipment standard may be erected. The regulation also requires that they are provided with a metal sheath, a metal enclosure or a fine mesh metallic grid. The grid, sheath or enclosure shall be connected to the protective conductor of the supply circuit, except where the protective measure of SELV is provided for the floor heating system.

The protective measure of electrical separation must not be used for electric floor heating systems in these locations.

12.10 Swimming pools and other basins

702.55.1 It is permitted to install an electric heating unit embedded in the floor, provided that it meets the requirements of Regulation 702.55.1.

12.11 Information for the user of the installation

Fig 753 A description of the heating system is required to be provided to the person ordering the work.

The description must contain at least the following information:

(a) description of the construction of the heating system, which must include the installation depth of the heating units;
(b) location diagram with information concerning:
  (i) the distribution of the heating circuits and their rated power;
  (ii) the position of the heating units in each room; and
  (iii) particularities that have been taken into account when installing the heating units, for example, heating-free areas, complementary heating zones, heating-free areas for fixing means penetrating into the covering material;
(c) data on the control equipment used, with relevant circuit diagrams and the dimensioned position of floor temperature and weather conditions sensors, if any; and
(d) data on the type of heating units and their maximum operating temperature.

The installer must inform the owner that the description of the heating system includes all necessary information, for example for repair work.

Instructions for use must be provided to the person ordering the work upon completion. One copy of the instructions for use must be permanently fixed in or near each relevant distribution board.

The instructions for use must include at least the following data:

(a) description of the heating system and its function;
(b) instructions on the operation of the heating installation in the first heating period in the case of a new building, e.g. for drying out;
(c) operation of the control equipment for the heating system in the dwelling area and the complementary heating zones, if any;
(d) information on restrictions on the placing of furniture or similar, including:
  (i) additional floor coverings. For example, carpets with a thickness of >10 mm may lead to higher floor temperatures which can adversely affect the performance of the heating system.
  (ii) pieces of furniture solidly covering the floor and/or built-in cupboards. These may be placed on heating-free areas.
  (iii) furniture, such as carpets, seating and rest furniture with pelmets, which in part do not solidly cover the floor. These may not be placed in complementary heating zones, if any.
(e) in the case of ceiling heating systems, restrictions regarding the height of furniture. Cupboards of room height may be placed only below the area of ceiling where no heating elements are installed.
(f) dimensioned position of complementary heating zones and placing areas.
(g) statement that, in the case of thermal floor, wall and ceiling heating systems, no fixing should be made into the floor, wall or ceiling respectively. Heating-free areas are excluded from this requirement. Alternatives must be given, where applicable.

Gardens (other than horticultural installations)

13

13.1 Introduction

Gardens and similar installations are not classified as special locations in BS 7671:2018; they must, therefore, comply with all the general requirements of the regulations that are applicable to outdoor circuits and equipment.

13.2 The risks

Gardens can present a number of potential risks where electrical installations are concerned. These risks may include:

▶ contact of persons with the general mass of Earth, possibly with bare feet;
▶ a frequently wet environment;
▶ the wearing of minimal clothing;
▶ gardening activity that may cause damage or disturbance to cables or equipment; and
▶ the insertion of spikes in the ground for securing marquees, inflatables etc.

In public gardens it should be recognized that electrical equipment, e.g. lighting, may be accessible to the general public.

13.3 Cables

13.3.1 Buried cables

522.8.10 Cables should be protected against foreseeable damage, either by earthed armouring or other suitable enclosure. Unprotected cables should not be buried directly in the ground, nor should they be clipped to wooden fences, etc., which may provide inadequate support and protection.

Problems arise when either ground levels are lowered so that cables have insufficient cover or when ground levels are raised so that cables that were not intended to be buried and are not suitable for burial become buried. Such problems can arise during the course of a project and the intended ground level should be formally ascertained before the cables are installed. It must be remembered that the layout of a garden can be changed totally within a few seasons and great care must be taken to route cables where they are not likely to be disturbed or damaged, e.g. around the edge of the plot and at sufficient depth.

Buried cable routes should be identified by local route markers and recorded on drawings. Cables should be buried at least 450 mm (preferably deeper) below the lowest local ground level, and a route marker tape laid along the cable route about 150 mm below the surface.

Note: Further information on cables buried in the ground is given in IET Guidance Note 1.

13.3.2 Cables exposed to sunlight

522.2.1
522.11.1
Cables should be shielded from prolonged exposure to direct sunlight, or be of a type suitable for such exposure. Cable with a black sheath is recommended. Generally, the ultraviolet light from the sun will affect plastics and the cable manufacturer's advice should be taken. However, cables must not be so enclosed that heat dissipation is inadequate.

13.3.3 Cables taken overhead

GN1
Cables taken overhead should have a suitable rigid support or a catenary wire, or be a type suitable for such installation. See Guidance Note 1, Appendix G.3 for more information.

13.4 Socket-outlets

411.3.3
Mobile equipment with a current rating not exceeding 32 A for use outdoors requires additional protection by means of a 30 mA RCD with characteristics in accordance with Regulation 415.1.1. Socket-outlets that can be expected to be used to supply this equipment outdoors will therefore require this additional protection, which can be provided by using an RCCB complying with BS 61008-1, an RCBO complying with BS EN 61009-1 or an SRCD complying with BS 7288.

Socket-outlets should be suitably placed to be convenient for their purpose, and of appropriate IP rating, e.g. IP44 or better if located outdoors.

13.5 Fixed equipment

Sect 559
714.1
Fixed equipment in the garden, such as permanently fixed garden lighting, should be securely fixed with all cables buried or supported away from the ground or paths. All-insulated Class II equipment is recommended where possible for increased safety. Decorative lighting, including 'festoon' type lighting, should be permanently fixed if used regularly.

Other fixed equipment, including pumps, etc., should be installed in accordance with the manufacturer's instructions and the general requirements of BS 7671:2018.

13.6 Ponds

Sect 702
In view of the risk of accidental or intentional immersion it is recommended that the same rules should be applied to garden ponds, especially larger ones, as are applied to swimming pools. Equipment (including cables) must be suitable for the purpose and of a suitable IP rating, or be installed in a suitable enclosure. Class II equipment should be utilized where possible. Cables should be installed in ducts or conduits built into the pond structure and not allowed to lie loose around the area. All connections must be made in robust, watertight junction boxes. Equipment to IP55 rating or better is recommended.

Pond lighting should meet the requirements of BS EN 60598-2-18, pumps BS EN 60335-2-41, and other equipment BS EN 60335-2-55.

13.7 Supplies to outbuildings within the garden

When supplying a detached garage or shed from a dwelling, the preferred method is to supply the outbuilding from a spare way in the consumer unit in the dwelling and to install a small distribution board in the outbuilding (see Figure 13.1). A suitably rated circuit-breaker should be fitted in a spare way in the dwelling's consumer unit and a submain cable installed to supply the consumer unit in the garage. The cable run from the house to the garage must be suitably protected, see section 13.3. The consumer unit in the garage should be fitted with circuit-breakers for the final circuits, such as 6 A for the lighting circuit and 16 A for the socket-outlet circuit.

13.7.1 Additions and alterations to an existing installation

132.16 Where work is being carried out to an existing installation Regulation 132.16 requires that the rating and condition of existing equipment, including that of the distributor, should be adequate for the additional load and that the existing earthing and bonding arrangements are also adequate.

13.7.2 Earthing and main protective equipotential bonding

411.3.1.2 Where an installation serves a detached building that contains extraneous-conductive-
Chap 54 parts, Regulation 411.3.1.2 requires main protective bonding conductors complying with Chapter 54 to connect to the main earthing terminal (in the house) extraneous-conductive-parts including:

(a) water installation pipes;
(b) gas installation pipes;
(c) other installation pipework and ducting;
(d) central heating and air conditioning systems; and
(e) exposed metallic structural parts of the building.

Table 54.8 For an outbuilding fed from a dwelling having a PME supply and where the outbuilding includes extraneous-conductive-parts, a main protective bonding conductor in accordance with Table 54.8 will be required (at least 10 mm² copper) to be run between the dwelling's main earthing terminal and the extraneous-conductive-parts in the outbuilding.

> **Note:** If there are no extraneous-conductive-parts in the outbuilding then protective equipotential bonding will not be required.

13.7.3 TT system

411.3.1.1 An alternative possibility is to make the small installation in the garage part of a TT system. A local earth electrode must be provided at the garage. To achieve protection by automatic disconnection of supply for both circuits in the garage an RCD must be employed. Main protective equipotential bonding will need to be provided at the garage connecting the metal water pipe and any other extraneous-conductive-parts to the earthing terminal in the small distribution board. An exposed-conductive-part connected to one means of earthing must not be simultaneously accessible with an exposed-conductive-part connected to another means of earthing (Regulation 411.3.1.1 refers). Where the installation in the garage is supplied by an armoured cable, the armour or any protective conductor in the cable must not be connected to, and must not be simultaneously accessible with, any exposed-conductive-parts in the outbuilding. The cable armour must be earthed at the supply end of the submain cable.

13.7.4 RCD protection

411.3.3 Regulation 411.3.3 requires additional protection by means of a 30 mA RCD to be provided for socket-outlets not exceeding 32 A. The exception in the regulation does not apply to dwellings. The regulation also requires additional protection by an RCD to be provided in circuits likely to supply mobile equipment for use outdoors.

RCD protection can be provided as follows:

▶ an RCBO in the dwelling's consumer unit for the garage supply; or
▶ an RCCB in the dwelling's consumer unit; or
▶ selecting a device that includes RCD protection as the main switch for the small two-way consumer unit in the garage; or
▶ protecting the socket-outlet circuit in the garage with an RCBO; or
▶ providing an SRCD (a socket-outlet incorporating RCD protection) for the socket-outlet in the garage.

13.7.5 Inspection, testing and certification

Part 6
GN 3 Inspection and testing must be performed to confirm the adequacy of the relevant parts of the existing installation that will support the changed requirements, the upgrading of the existing installation necessary to support the addition or alteration, and the addition or alteration itself. The requirements for initial verification are contained in Chapter 64 of Part 6 of BS 7671:2018 and further information on the requirements for inspection and testing is given in Guidance Note 3: *Inspection & Testing*. Compliance with BS 7671 must be verified for every addition or alteration. Requirements for certification and reporting in respect of electrical installations are also given in Part 6. An Electrical Installation Certificate must be provided to the person ordering the work giving details of the extent of the installation covered by the certificate, together with a record of the inspection and the results of the testing.

▼ **Figure 13.1** Outbuilding supply taken from a spare way in the consumer unit

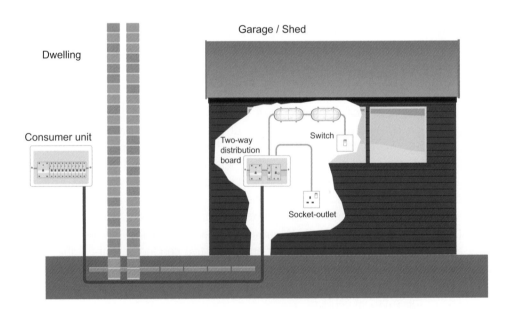

13.8 Supplies to hot tubs

Persons using hot tubs are particularly at risk due to:

(a) lack of clothing, particularly footwear;
(b) presence of water which reduces contact resistance;
(c) immersion in water, which reduces total body resistance;
(d) ready availability of earthed metalwork;
(e) increased contact area; and
(f) contact with ground where outdoors.

There are no specific requirements in BS 7671 for electrical installations associated with hot tubs. However, supplies to hot tubs should be protected by a 30 mA RCD.

Sect 701 Where a hot tub is located in a room other than a bathroom, for example, a garden room/shed/etc., it is recommended that the requirements of Section 701 should be applied in full. The requirements are given in Chapter 1 of this guide.

Sect 702 Where a hot tub is located outdoors in the open air it is recommended that the requirements of Section 702 should be applied in full. The requirements are given in Chapter 2 of this guide.

Persons involved in the design of the electrical installation for a hot tub should consult the product standard BS EN 60335-2-60: *Specification for safety of household and similar appliances: Particular requirements for whirlpool baths and whirlpool spas.*

13

Mobile or transportable units

14

14.1 Introduction

Sect 717   There are no significant changes to Section 717 introduced by the 18th Edition.

General

Installations of mobile and transportable units are included in Section 717 of BS 7671. These introduce several considerations that are not normally present in installations designed for buildings. In buildings, the source of supply is clearly defined and the earthing arrangements are understood. There are many ways in which mobile, etc. units are used; on each occasion they might receive a supply from a different fixed installation or generator, each of which will have unique electrical characteristics. There are also many different kinds of unit, designed to serve a variety of different activities. The installations will range from simple to very complex and the people operating the units might be ordinary people or those with some technical understanding, though not necessarily electrically skilled persons. Units might operate singly with an on-board generator or from an external supply but otherwise completely separate from any other electrical system. They might operate in groups of two or more units; these might be parked close to each other and use the same supplies or be supplied from different sources. Operation in groups of two or more is likely to occur in the entertainment, broadcast and possibly the emergency services; in such case it is likely that units will be interconnected through signal/data cables that are likely to be earth referenced.

It is therefore important for the designer to check that all the ways in which the unit will be used are fully understood, including the knowledge and training of those using or operating the unit, before commencing a design.

The figures in this chapter are a guide to the principles that should be adopted for installations in mobile and transportable units; they should not be taken as the exact way that installations in units should be arranged. They are only examples. Figures 14.1 (no 14.2) to 14.7 are based upon those shown in HD 60364-7-717.

14.2 Scope

Sect 717.1   The particular requirements of Section 717 apply to AC and DC installations for mobile or transportable units.

The term 'unit' is intended to mean a vehicle and/or mobile or transportable structure in which all or part of an electrical installation is contained.

Units are either one of two types:

(a) mobile, e.g. vehicles (self-propelled or towed); or
(b) transportable, e.g. containers or cabins.

Examples of the units include technical and facilities vehicles for the entertainment industry, medical or health screening services, welfare units, promotion & demonstration, firefighting, workshops, offices, transportable catering units, etc.

The requirements are not applicable to:

(a) generating sets (Section 551 of BS 7671);
(b) marinas and pleasure craft (Section 709);
(c) mobile machinery in accordance with BS EN 60204-1;
(d) caravans (Section 721);
(e) traction equipment of electric vehicles; or
(f) electrical equipment required by a vehicle to allow it to be driven safely or used on the highway.

The guidance given here is in addition to the general requirements of BS 7671. Where other special locations, such as rooms containing showers or medical locations, form part of a mobile or transportable unit, the particular requirements for those installations should also be taken into consideration. In this regard, reference should be made to Part 7 of BS 7671:2018 and the other relevant chapters of this Guidance Note.

Note: Guidance on temporary electrical systems for events, entertainment and related purposes is given in BS 7909. Reference to BS 7909 or HSE document OC 482/2 may be helpful when considering supplies to units.

14.3 The risks

The risks associated with mobile and transportable units vary due to differences in design and use, but typically arise from:

(a) risk of loss of connection to earth, due to the use of temporary cable connections and long supply cable runs, the repeated use of cable connectors, which may give rise to 'wear and tear', and the potential for mechanical damage to these parts;
(b) risks arising from the connection to different national, local electricity distribution networks and generators, where unfamiliar supply characteristics and earthing arrangements are found;
(c) impracticality of establishing an equipotential zone external to the unit;
(d) open-circuit faults of the PEN conductor of PME supplies raising the potential of all metalwork (including that of the unit) to dangerous levels;
(e) risk of shock arising from high functional currents flowing in protective conductors – usually where the unit contains substantial amounts of electronics or communications equipment; and
(f) vibration while the vehicle or trailer is in motion, or while a transportable unit is being moved – causing faults within the unit installation.

Particular requirements to reduce these risks include:

(a) designing the installation to be safe and suitable for the service to be provided and the abilities of those who will use and operate the unit.

(b) checking the suitability of the electricity supply before connecting the unit.

(c) accessible conductive parts of the unit to be connected through the main protective equipotential bonding to the main earthing terminal within the unit.

(d) a regime of regular inspection and testing of connecting cables and their couplers, supported by a logbook system of record keeping.

(e) the use of flexible cables.

(f) the use of RCDs.

(g) clear and unambiguous labelling of units, indicating types and characteristics of supply that may be connected. Instruction manuals, circuit diagrams and/or electrical installation certificates should give this information in a form that will be clearly understood by those people using or operating the unit.

(h) particular attention paid to the maintenance and periodic inspection of installations.

14.4 Supplies

Regarding mobile and transportable units, the term 'supply' needs to be carefully considered as many variations might exist, for instance:

(a) ice cream, fish and chip and similar food dispensing units might always connect to their own on-board or external generator or supply, the characteristics of which are known.

(b) mobile X-ray and similar medical units might only attend specific surgeries or similar premises where specific supply connection points have been installed for their use.

(c) some sports venues or places of entertainment that expect broadcast and similar units might have supply connection points that can provide for such units.

(d) many units seek to use supplies wherever they provide their service. These supplies might be from any installed system at a premises or from a generator; in this case the supplies could have a very wide range of characteristics.

(e) at larger sporting or entertainment events it is not unusual for multiple units to be present and for there to be two or more different sources of supply.

Because of the enormous variation in the characteristics of the supplies that might be encountered it is important that designers fully understand the service that the unit will provide, the range of supplies that a particular unit will use and the knowledge of electrical systems that the users or operators of the unit have. These factors will affect the design of the unit's electrical installation.

Reference to BS 7909 or HSE document OC 482/2 may be helpful when considering supplies.

14.4.1 General

717.313 One or more of the following methods must be used to supply a unit:

(a) connection to a low voltage generating set, located inside the unit, in accordance with Section 551 (see Figure 14.1);

(b) connection to a low voltage electrical supply external to the unit, in which the protective measures are effective (see Figure 14.3), the supply derived from either a fixed electrical installation or a generating set in accordance with Section 551; and

(c) connection to a low voltage electrical supply external to the unit, and where internal protective measures are provided by the use of simple separation, in accordance with Section 413, showing alternative forms of fault protection within the unit (see Figures 14.4 to 14.7).

The following notes enlarge upon details of the supplies outlined in (a), (b) and (c) immediately above.

Note 1: In cases (a), (b) and (c), an earth electrode may be provided where supplies are used external to the vehicle (see Regulation 717.411.4).

Note 2: In case (c), an earth electrode may be necessary in Figure 14.4 for protective purposes (see Regulation 717.411.6.2(ii)).

Note 3: Simple separation or electrical separation is appropriate for any combination for the following reasons:

(i) it allows connection to a supply with any earthing arrangement without introducing the risks of electric shock associated with an open-circuit PEN in a PME distribution. See Figs 14.4 to 14.7.

(ii) it allows the internal installation to be arranged as IT, see Figs 14.4, 14.5 and 14.7 (see also 14.4.3). Or arranged as TN-S, see Fig 14.6, which is probably the easiest internal earthing arrangement to deal with.

(iii) where high residual currents are expected these are retained within the unit's installation and do not affect the supply. Such currents can be expected where a unit is equipped with information technology, broadcast, communications or similar equipment.

(iv) where a reduction of electromagnetic disturbances is necessary.

(v) if the supply to the unit comes from alternative supply systems (as is the case in disaster management).

The source, means of connection or separation may be within the unit.

Note 4: There are potential hazards when a unit is about to be moved if all equipment or facilities that are used in operation whilst static have not been correctly de-rigged, stowed and secured. Such things include:

(a) connection of external electrical supplies and signal cables, etc.;

(b) connection of external water, waste, gas and similar facilities;

(c) closure or stowing of service or access hatches and doors, stairs, ladders, guard and handrails, etc.;

(d) stowing of aerials, satellite dishes and rigged lighting, etc.; and

(e) closure or stowing of all awnings, expanding sides and other facilities that increase the external dimensions of the unit when static and in use.

It is recommended that where these facilities exist the unit is equipped with an electrical interlock or other means, as might be appropriate, that warns, alarms, or prevents movement in order to reduce the risk. This might involve the electrical system required by the vehicle for motion and use on the highway.

Note 5: For the purpose of this section, power inverters or frequency convertors supplied from the unit's electrical system or an auxiliary system driven by the unit's prime mover or an on-board battery are also considered as generating sets.

Power inverters or frequency convertors used in this way shall include electrical separation where both the DC supply and the AC neutral point are connected to the main earthing terminal. See also 14.4.4, where the use of frequency convertors in place of 50 Hz transformers is considered.

Note 6: In cases (a) and (c) (methods of supply – clause 14.4.1) applicable to Figs 14.1 and 14.6, the earthing arrangement of the installation in the unit is seen as TN-S, whether or not an earth electrode is deployed.

Note 7: In case (b) (methods of supply - clause 14.4.1 – see Fig 14.3) the installation in the unit will take on the earthing arrangement of the source of the supply to which the unit is connected, which might be TT, TN-S or TN-C-S.

14.4.2 TN system

717.411.4 A PME earthing facility shall not be used as the means of earthing for an installation falling within the scope of Section 717, except where:

(a) the installation is continuously under the supervision of a skilled or instructed person competent in such work; and

(b) the suitability and effectiveness of the means of earthing has been confirmed before the connection is made.

Note: Regulation 717.411.4 recognizes the possibility (though rare) of an open-circuit PEN conductor occurring in a PME distribution. Where a protective earth for a unit is derived from a PME source (a TN-C-S supply) there is the possibility of receiving an electric shock if an open-circuit PEN should occur. In this circumstance a person would receive an electric shock when outdoors and outside the unit and touching the conductive structure of the unit. Hence the general prevention of the use of a PME earthing facility; this is specifically applicable to installations following Fig 14.3, or if the use of a protective earth from an unknown supply is being considered.

There are many installations in premises that are TN-C-S in form and ordinary persons will be most unlikely to understand and fulfil the requirements set out in (a) and (b) above. It is therefore recommended that the arrangements of Fig 14.6 are followed wherever an ordinary person is required to use, operate or connect a unit to a supply with unknown or uncertain earthing arrangements.

Note: There are people who use or operate units with an installation related to Fig 14.3 or 14.6 and are sufficiently skilled and competent to understand and fulfil the requirements of (a) and (b) above. In this case, in the UK there is a variation in design that is based on Fig 14.6 but has an incoming protective earth connection similar to Fig 14.3. The supply intake to Fig 14.6 is either a three-phase and neutral intake to a star/star or three-phase intake to a delta/star transformer or single-phase, but includes a protective conductor with the live conductors. This protective conductor is connected to the main earthing terminal and contains a removable earth link accessible with the use of a tool. This permits the skilled or instructed person competent in such things to decide how the earthing arrangements of the on-board installation shall be set up for best results in the conditions of use with the supplies available.

14.4.3 IT system

717.411.6.2 An IT system can be provided by:

(a) an isolating transformer or a low voltage generating set, with an insulation monitoring device (IMD) or an insulation fault location system, both without automatic disconnection of the supply in case of the first fault and without a need of connection to an earthing installation (see Figure 14.7); the second fault shall be automatically disconnected by overcurrent protective devices according to Regulation 411.6.4; or

(b) a transformer providing simple separation, e.g.in accordance with BS EN 61558-1, with an RCD and an earth electrode installed to provide automatic disconnection in the case of failure in the transformer providing the simple separation (see Figure 14.4).

Note: In the UK systems with IT earthing arrangements are typically used in medical locations requiring security and safety of supplies to equipment involved with life-support for patients; for this reason some mobile and transportable units might have internal IT installations. Such medical IT installations should follow the requirements of Section 710 of BS 7671:2018.

IT installations might be appropriate for other units if this degree of security and safety of supply is required. The use of IT is fairly unusual in the UK, though it is used widely in certain European and other countries and needs to be taken into account if units are for service outside the UK.

In the UK there are some units used by the communications industries that have IT installations. Most units used by these industries have installations that are TN-S (as a version of Fig 14.1 or 14.6) or take on the earthing arrangements from their source of supply (as Fig 14.3). Where units with IT installations are likely to be interconnected via earth-referenced signal cables with other units used by these or related industries, care must therefore be taken that the interconnection of protective conductors of different systems does not take place. The typical solution is for the unit with an IT installation to have signal isolation devices at each signal interconnection point on that unit, or for interconnections to be made using fibre optics.

14.4.4 Supply intake transformers

The intake arrangements in Figs 14.4 to 14.7 all show a delta/star three-phase transformer; they could be star/star transformers or they could be for a single-phase supply. A transformer makes it possible to accept a supply with any earthing arrangement. This transformer might be of the general type operating at 50 Hz or might be of the active or frequency converting type; whichever type is selected it must at least provide electrical separation. Active or frequency converting transformers of up to about 15 kVA might be used in place of 50 Hz transformers as they can be smaller and lighter. Where active or frequency converting transformers are used the manufacturer's instructions should be followed regarding earthing at the input and output and the provision of protection in the circuits supplied by the output.

Wherever a unit's installation includes electronic or digital equipment, for instance, that used in entertainment, broadcast, communications and emergency units etc., high residual currents can be expected in protective conductors. The transformer ensures that these are retained within the unit's installation and do not affect the source of supply, where they might trip upstream circuit protective devices.

Installations in units with electronic and digital equipment of the types outlined are likely to experience high levels of harmonic currents due to the switching cycles of the equipment being supplied. These harmonics can cause overheating in transformers and can overload the neutral in three-phase systems.

There are two important characteristics that should be specified when selecting intake transformers:

(a) the ability to handle the harmonics expected in the transformer of the type selected; and

(b) the transformer should have a low inrush current, otherwise trouble might be experienced with tripping of upstream circuit protective devices every time the supply is switched on.

Transformers operating at 50 Hz are bulky and heavy items and can be difficult to position satisfactorily in a unit.

The frequency converting types of similar rating might offer a saving in size and weight. Space and load distribution in a vehicle are important parameters, making the selection of the right transformer and placing it in the unit for best physical and electrical results an important consideration.

14.4.5 Units with a second supply

Although not outlined in any of the figures, it is not unusual for some units to have more than one supply. A typical situation is as follows:

A unit with an on-board generator arranged as Fig 14.1 will have a limited quantity of fuel or stored energy, so those operating the unit (for instance, a communications unit) will seek the security of a second supply, which might be provided by connection to an external generator or a fixed installation at a premises. Where this facility is provided, selection of supply is controlled via an on-board three-position switch that switches all live conductors and offers 'on-board generator' – 'off' – 'external supply'. The intake of the external supply is arranged via a transformer as Fig 14.6 with the third position of the switch taking the output from the transformer. The transformer used in this case is often a frequency converting type. The output of the transformer is also connected to the main earthing terminal; a protective conductor associated with the external source is not connected to the main earthing terminal. This arrangement means that the earthing arrangement of the unit's installation remains the same (as TN-S), no matter which supply is used.

▼ **Figure 14.1** (Fig 717.1 of BS 7671) – An example of a connection to a low voltage generating set located inside the unit, with or without an earth electrode

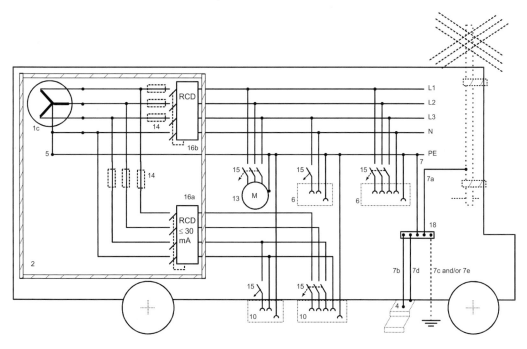

▼ **Figure 14.3** (Fig 717.3 of BS 7671) – An example of a connection to a low voltage electrical supply external to the unit in which the protective measures are effective, the supply derived from either a fixed electrical installation or a generating set, with or without an earth electrode at the unit

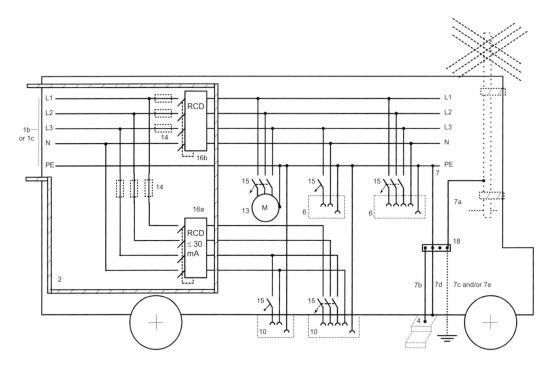

▼ **Figure 14.4** (Fig 717.4 of BS 7671) – An example of a connection to a low voltage electrical supply, external to the unit, derived from either a fixed electrical installation or a generating set with any type of earthing arrangement using simple separation and an internal IT system, with an earth electrode

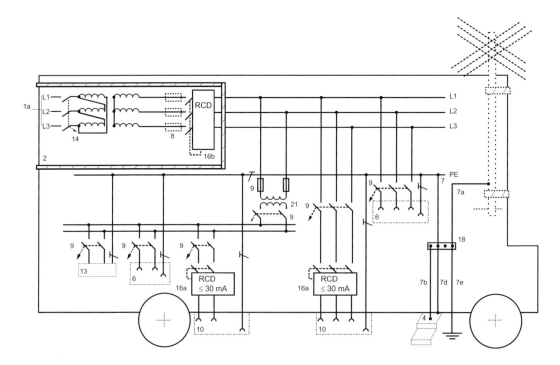

▼ **Figure 14.5** (Fig 717.5 of BS 7671) – An example of a connection to a low voltage
electrical supply, external to the unit, using simple separation and an internal
IT system with an insulation monitoring device and automatic disconnection
of supply on the occurrence of a first fault, with earth electrode

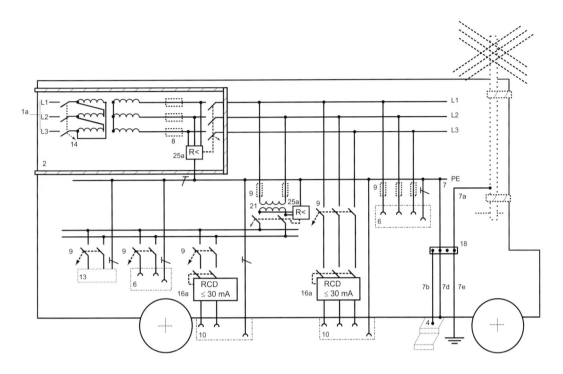

▼ **Figure 14.6** (Fig 717.6 of BS 7671) – An example of a connection to a low voltage electrical
supply, external to the unit, with any type of earthing arrangement using an
internal TN system with simple separation

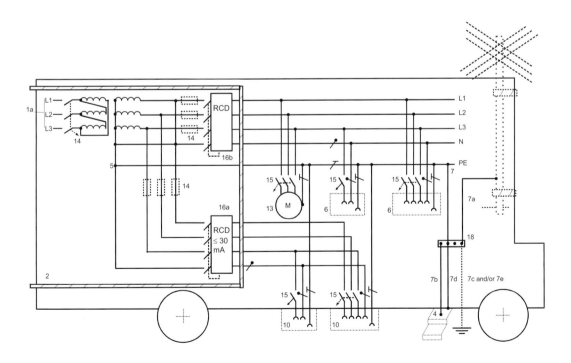

▼ **Figure 14.7** (Fig 717.7) – An example of a connection to a low voltage electrical supply, external to the unit, with any type of earthing arrangement using simple separation with an IT system with automatic disconnection on the occurrence of a second fault

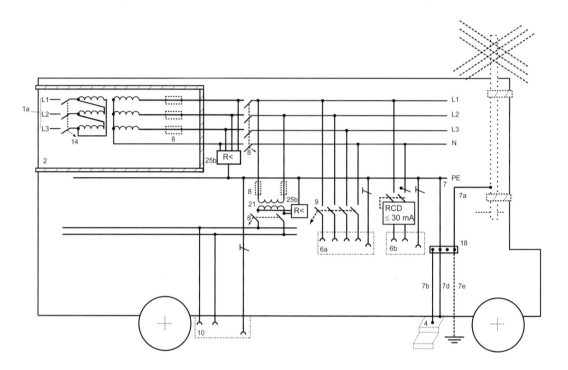

Key to Figures 14.1 to 14.7

1a Connection of the unit to any external supply through an on-board transformer with simple separation

1b Connection of the unit to a supply in which the protective measures are effective

1c Connection to an LV generator set in accordance with Section 551

2 Class II or equivalent enclosure up to the first protective device providing automatic disconnection of supply

4 Conductive external staircase, if any

5 Connection of the neutral point (or, if not available, a line conductor) to the conductive structure of the unit

6 Socket-outlets for use exclusively within the unit

6a Socket-outlets for use exclusively within the unit for reasons of continuity of supply in the event of first fault

6b Socket-outlets for general use if explicitly required (operation of the RCD in the event of first fault cannot be excluded)

7 Protective equipotential bonding in accordance with Regulation 717.411.3.1.2

7a to an antenna pole, if any

7b to the conductive external stairs, if any, in contact with the ground

7c to a functional earth electrode, if required

7d to the conductive structure of the unit

7e to an earth electrode for protective purposes, if required

8 Protective devices, if required, for overcurrent and/or for protection by disconnection of supply in case of a second fault

9 Protective devices for overcurrent and for automatic disconnection of supply in case of a second fault

10 Socket-outlets for current-using equipment for use outside the unit

13 Current-using equipment for use exclusively within the unit

14 Overcurrent protective device, if required

15 Overcurrent protective device

16a Residual current device (RCD) having the characteristics specified in Regulation 415.1.1 for protection by automatic disconnection of supply for circuits of equipment for use outside the unit

16b RCD for protection by automatic disconnection of supply for circuits of equipment for use inside the unit: see Regulations 411.4.4 and 411.5.3. Where an internal IT system is installed, see also Regulation 411.6.4

18 Main earthing terminal or bar

21 Transformer with at least simple separation e.g. 230 V current-using equipment

25a Insulation monitoring device providing disconnection on the first fault, see Figure 14.5

25b Insulation monitoring device or insulation fault location system including monitoring of the N conductor if distributed (disconnection only in the event of second fault), see Figure 14.7.

14.5 Protection against electric shock

14.5.1 General

717.410.3.5 The protective measures of obstacles and placing out of reach (Section 417) shall not be used.

717.410.3.6 The protective measure of non-conducting location (Regulation 418.1) shall not be used and earth-free local equipotential bonding (Regulation 418.2) is not recommended.

14.5.2 Protective measure: electrical separation

717.413 Regulation 717.413 gives requirements for the protective measure of electrical separation.

Electrical separation by transformer providing simple separation, e.g. complying with BS EN 61558-1, in accordance with Regulation 413.1.3 shall be used only where:

(a) an insulation monitoring device is installed to provide automatic disconnection of the supply in the case of a first fault between live parts and the conductive structure of the unit (see Figure 14.5); or

(b) an RCD and an earth electrode are installed to provide automatic disconnection in the case of failure in the transformer providing the electrical separation (see Figure 14.4). Each socket-outlet intended to supply current-using equipment outside the unit shall be protected individually by an RCD having the characteristics specified in Regulation 415.1.1.

14.5.3 Additional protection

717.415 Every socket-outlet intended to supply current-using equipment outside the unit shall be provided with additional protection by an RCD with a rated residual operating current not exceeding 30 mA, with the exception of socket-outlets which are supplied from circuits with protection by:

(a) SELV; or

(b) PELV; or

(c) electrical separation, with an insulation monitoring device; see Regulation 717.413(i).

Note: IET Guidance Note 5 gives guidance on the protective measure of electrical separation.

14.6 Protective equipotential bonding

717.411.3.1.2   Accessible conductive parts of the unit, such as the conductive structure of the unit, shall be connected through the main protective bonding conductors to the main earthing terminal within the unit.

The protective bonding conductors shall be finely stranded. Cable types H05V-K and H07V-K to BS EN 50525-2-31 are considered appropriate.

14.7 Wiring systems external to the mobile or transportable unit

717.52.1   Where the supply to the mobile or transportable unit is provided by means of a plug and socket-outlet, flexible cables in accordance with H07RN-F (BS EN 50525-2-21) or cables of equivalent design, having a minimum cross-sectional area of 2.5 mm^2 copper, shall be used for connecting the unit to the supply.

The flexible cable shall enter the unit by an insulating inlet in such a way as to minimize the possibility of any insulation damage or fault that might energize the exposed-conductive-parts of the unit.

14.8 Wiring systems inside the mobile or transportable unit

717.52.2   The wiring system shall be installed using one or more of the following:

(a) Unsheathed flexible cable with thermoplastic or thermosetting insulation to BS EN 50525-2-31, -3-31 or BS EN 50525-3-41 installed in conduit in accordance with the appropriate part of the BS EN 61386 series or in trunking or ducting in accordance with the appropriate part of the BS EN 50085 series.

(b) Sheathed flexible cable with thermoplastic or thermosetting insulation to BS EN 50525-2-11, -2-21, -3-11 or -3-21, if precautionary measures are taken such that no mechanical damage is likely to occur due to any sharp-edged parts or abrasion.

All cables shall, as a minimum, meet the requirements of BS EN 60332-1-2.

Conduits shall comply with BS EN 61386-21, -22 or -23.

14.9 Proximity to non-electrical services

717.528.3.4   No electrical equipment, including wiring systems, except ELV equipment for gas supply control, shall be installed in any gas cylinder storage compartment.

ELV cables and electrical equipment may only be installed within the LPG cylinder compartment if the installation serves the operation of the gas cylinder (e.g. indication of empty gas cylinder), or is for use within the compartment.

Such electrical installations and components shall be constructed and installed so that they are not a potential source of ignition.

Where cables have to run through such a compartment, they shall be protected against mechanical damage by installation within a conduit system complying with the appropriate part of the BS EN 61386 series or within a ducting system complying with the appropriate part of the BS EN 50085 series.

Where installed, this conduit or ducting system shall be able to withstand an impact equivalent to AG3 (high severity) without visible physical damage.

14.10 Identification and notices

717.514 A permanent notice of durable material and likely to remain easily legible throughout the life of the installation shall be fixed to the unit in a prominent position, preferably adjacent to the supply inlet connector. The notice should state in clear and unambiguous terms the following information:

(a) the types of supply which may be connected to the unit and any limitations on use imposed by the designer;
(b) the voltage rating of the unit;
(c) the number of supplies, phases and their configuration;
(d) the on-board earthing arrangement; and
(e) the maximum power requirement of the unit.

14.11 Other equipment

717.55 Where the means of connection is a plug and socket-outlet, mounted, accessed or
717.55.1 used outside the unit and used to connect the unit to the supply, or supply other equipment, it shall comply with the appropriate parts of the BS EN 60309-2 series and shall meet with the following requirements:

(a) plugs shall have an enclosure of insulating material;
(b) connecting devices, plugs and socket-outlets, with an enclosure as necessary, shall afford a degree of protection of at least IP44 rating when in use or connected and protection of at least IP55 rating when not connected, e.g. when the unit is in transit; and
(c) the inlet (with 'male' contacts) shall be situated on the unit.

717.55.3 **Where a generating set provides a supply as a switched alternative to the normal supply to the installation**

717.551.6 Regulation 717.551.6 prohibits units with different power supply systems being interconnected. This regulation also prohibits different earthing systems being interconnected, except where special precautions have been taken. This reinforces the general rules in earlier parts of the Regulations.

Live conductors from different power supplies shall not be interconnected.

Protective conductors, including functional earthing conductors, from different earthing systems shall only be interconnected where suitable precautions have been taken into account; see also Regulation 542.1.3.3.

Plugs and socket-outlets shall comply with the appropriate parts of the BS EN 60309 series, except those intended for special equipment, such as broadcasting equipment where combined connectors for information signals and power supply are used.

Note: In the broadcasting, communication, entertainment industries etc. units are frequently interconnected with earth-referenced signal cables for the transfer of data. Such earth-referenced signal cables might cause the interconnection of earthing systems in units provided with different supplies. The designer of the arrangements for signal transfer between units needs to consider the use of signal isolation devices at the points for interconnection; the use of fibre optics might be appropriate in this situation.

717.551.7.2 **Where a generating set may operate in parallel with other sources including systems for distribution of electricity to the public**

Regulation 717.551.7.2 gives additional requirements for installations where the generating set may operate in parallel with other sources. This again reinforces the general rules in earlier parts of the Regulations.

A generating set used as an additional source of supply in parallel with another source shall only be connected on the supply side of all the protective devices for the final circuits of the installation.

Protective conductors, including functional earthing conductors, from different earthing systems shall only be interconnected where suitable precautions have been taken into account; see also Regulation 542.1.3.3.

Note: The contents of the note above also apply here.

14.12 Inspection, testing and certification

Chap 64   The installations in mobile and transportable units should, during erection and on completion before being put into service, be inspected and tested to verify, as far as reasonably practicable, that the requirements of the Regulations have been met and be provided with Electrical Installation Certificates etc. as required by Part 6 of BS 7671. The verification, inspection and testing work should be based upon the design parameters of the particular unit. Much of this work will follow a practice applicable to installations in premises, but some will require careful consideration due to the variety of supplies that might be encountered by a unit and the design of that unit. The following will require particular consideration and should be specified in the design:

(a) the supply or types of supply that can be used (TT, TN-S, TN-C-S, IT, or generator);

(b) the supply voltage, frequency, and maximum current;

(c) the live loop impedance of the source of supply, or range of live loop impedance that the source of supply should fall within and any conditions that should be met;

(d) the earth fault loop impedance of the supply, or range of earth fault loop impedance that the source of supply should fall within and any conditions that should be met;

(e) the earthing arrangement of the internal installation; and

(f) the use or otherwise of an earth electrode.

Entering the information applicable to the above data on a standard Electrical Installation Certificate presents some difficulties, as no model electrical installation certificate exists that is entirely suitable for mobile and transportable units. It is recommended that a standard electrical installation certificate is used, with details and explanations given on a separate sheet attached to it. It is usual for the mobile unit to be identified by a serial number or vehicle registration number, with due reference given to the owner and the owner's address.

It is also recommended that circuit diagrams and user manuals that might be applicable to a unit should contain in clearly understandable language any specific design criteria that are important to the operation of the unit or which might limit its use.

14.13 Routine maintenance and testing

The service duty of mobile and transportable units will vary with the type of unit and type of use (e.g. owner-operator or hire) but it is likely that frequent connecting and disconnecting combined with transporting will amount to the equivalent of rough service life. Frequency of use should therefore be an important factor in determining inspection and testing intervals. It is recommended that a visual inspection is carried out on the connecting cable and all plugs and socket-outlets before each and every transported use of the unit. The results of the visual inspection should be entered in a logbook as a permanent record of the condition of the electrical equipment. No repairs or extensions are acceptable on the external cable system and this should be replaced in its entirety if there are signs of damage or wear and tear (abrasion).

Chap 65 As a minimum, the unit electrical system should be inspected and tested annually, a report obtained on its condition and the necessary maintenance work, if any, implemented before the unit is next used. The recommendations of Part 6 of BS 7671:2018 should be followed in this regard, together with the specific guidance given in IET Guidance Note 3. If the unit duty is considered to be arduous, the inspection intervals should be reduced to cater for the particular conditions experienced.

All RCDs should be tested regularly by operating the test button, and periodically by a proprietary test instrument to verify that their operation remains within the parameters of the relevant British Standard.

All tests should be tabulated for record purposes. The necessary forms required by Part 6 of BS 7671 must be provided to the person ordering the work, by the contractor or persons carrying out the inspection and tests.

Guidance Note 7: Special Locations
© The Institution of Engineering and Technology

Small-scale embedded generators (SSEG)

15

A wind generator. Courtesy Dr Sung

15.1 Introduction and the law

There are no significant changes regarding generating sets introduced by the 18th Edition.

Sect 551 Low voltage generating sets such as small-scale embedded generators are not classed as special locations or installations under BS 7671:2018. There are, however, regulations related to this type of installation in Section 551 'Low voltage generating sets'.

15.1.1 The Electricity Safety, Quality and Continuity Regulations 2002

Over time the use of small-scale embedded generators (SSEG) is likely to become widespread. The ESQCR 2002 exempt sources of energy with an electrical output not exceeding 16 A per phase at low voltage (230 V) from Regulations 22(1)(b) and 22(1) (d) of the Regulations, and this chapter covers such sources of energy.

In addition to the electrical output limit of 16 A per phase (sub-paragraph (a) of Regulation 22(2)), other requirements for such small generators are:

> *(b) the source of energy is configured to disconnect itself electrically from the parallel connection when the distributor's equipment disconnects the supply of electricity to the consumer's installation; and*
>
> *(c) the person installing the source of energy ensures that the distributor is advised of the intention to use the source of energy in parallel with the network before, or at the time of, commissioning the source.*

The requirements that still need to be met in Regulation 22(1) are:

> *. . . No person shall install or operate a source of energy which may be connected in parallel with a distributor's network unless he –*
>
> *(a) has the necessary and appropriate equipment to prevent danger or interference with that network or with the supply to consumers so far as is reasonably practicable;*
>
> *(c) where the source of energy is part of a low voltage consumer's installation, complies with the provisions of British Standard Requirements [meaning, by definition, BS 7671].*

15.1.2 BS 7671:2018 Section 551 – Low voltage generating sets

551.2 Section 551 includes further requirements in Regulation 551.2 to ensure the safe connection of low voltage generating sets including small-scale embedded generators.

551.4.2 Regulation 551.4.2 relates to the use of RCDs. This regulation states:

> *The generating set shall be connected so that any provision within the installation for protection by RCDs in accordance with Chapter 41 remains effective for every intended combination of sources of supply.*

551.1 Note A note to Regulation 551.1 informs that the procedure for connecting generating sets up to 16 A in parallel with the public supply is given in ESQCR. Requirements of the distributor for the connection of units rated up to 16 A are given in BS EN 50438 *Requirements for the connection of micro-generators in parallel with public low-voltage distribution networks*.

551.7 For generating sets above 16 A the requirements of the distributor must be ascertained. The 18th Edition recognizes that there are two connection options:

(a) connection into a separate dedicated circuit; and

(b) connection into an existing final circuit.

Connection into a dedicated circuit is preferred.

Regulations 551.7.1 and 551.7.2 set out the requirements for the two options. The regulations require that a generating set used as an additional source of supply in parallel with another source shall either be installed on the supply side of all protective devices for the final circuits of the installation (connection into a separate dedicated circuit) or, if connected on the load side of all protective devices for the final circuits, must fulfil a number of additional requirements as follows:

(a) the current-carrying capacity of the final circuit conductors shall be greater than, or equal to, the rated current of the protective device plus the rated output of the generating set;

(b) a generating set shall not be connected to a final circuit by means of a plug and socket-outlet;

(c) an RCD providing additional protection of the final circuit in accordance with Regulation 415.1 shall disconnect all live conductors, including the neutral conductor;

(d) the line and neutral conductors of the final circuit and of the generating set shall not be connected to Earth; and

(e) unless the device providing automatic disconnection of the final circuit in accordance with Regulation 411.3.2 disconnects the line and neutral conductors, it shall be verified that the combination of the disconnection time of the protective device for the final circuit and the time taken for the output voltage of the generating set to reduce to 50 V or less is not greater than the disconnection time required by Regulation 411.3.2 for the final circuit.

The generator equipment should be type-tested and approved by a recognized body.

15.2 Engineering Recommendations

Note: A new Engineering Recommendation G98 was published on the 16 May 2018.

G98 supersedes Engineering Recommendation G83 from 27 April 2019. Until then, persons can use either G83 or G98. On 27 April 2019 G83 will be withdrawn and all new generation from that date must comply with the requirements of G98.

Engineering Recommendation G98 contains requirements for the connection of Fully Type Tested Micro-generators (up to and including 16 A per phase) in parallel with public Low Voltage Distribution Networks on or after 27 April 2019.

Guidance from Engineering Recommendation G83 is given below.

To assist network operators and installers, the Energy Networks Association prepared Engineering Recommendation G83: *Recommendations for the connection of small-scale embedded generators (up to 16 A per phase at low voltage) in parallel with public low voltage distribution networks*. The guidance given in this chapter is based on some of the requirements of the Engineering Recommendation as they would apply to persons responsible for electrically connecting such generators.

Engineering Recommendation G83 is for all small-scale embedded generator (SSEG) installations with an output up to and including 16 A per phase, single or polyphase, 230/400 V AC This includes:

(a) domestic combined heat and power;

(b) hydro;

(c) wind power;

(d) photovoltaic; and

(e) fuel cells.

Requirements for solar photovoltaic (PV) installations are found in Chapter 10 of this Guidance Note.

The Engineering Recommendation describes two types of connection procedure for the connection of an SSEG unit(s) in parallel with the public low voltage distribution network.

The first is a Single Premises Connection Procedure. The installer firstly has to make the distribution network operator (DNO) aware of the SSEG installation at or before the time of commissioning, in accordance with the ESQCR. The installer then has to provide the DNO with all the necessary information on the installation within 28 days of the SSEG unit being commissioned.

The second is a Multiple Premises Connection Procedure. The installer would be expected to discuss the project with the local DNO prior to installation, who will assess the impact of the connections on the network and provide the appropriate connection conditions. The installer then has to provide the DNO with all the necessary information on the installation within 28 days of commissioning.

Engineering Recommendation G83 incorporates forms, which define the information required by a public DNO for a small-scale embedded generator that is connected in parallel with a public low voltage distribution network. Supply of information in this form, for a suitably type-tested unit, is intended to satisfy the legal requirements of the DNO and hence will satisfy the legal requirements of the ESQCR.

15.3 Installation and wiring

The installation that connects the embedded generator to the supply terminals must comply with BS 7671.

A suitably rated overcurrent protective device shall protect the wiring between the electricity supply terminals and the embedded generator.

The SSEG must be connected directly to a local isolating switch. For single-phase machines the phase and neutral are to be isolated and for polyphase machines all phases and neutral are to be isolated. In all instances the switch, which must be manual, shall be capable of being secured in the 'off' isolating position. The switch is to be located in an accessible position in the customer's installation.

15.4 Means of isolation

551.7.6   Regulation 551.7.6 states the requirements for isolation:

> **551.7.6** *Means shall be provided to enable the generating set to be isolated from the system for distribution of electricity to the public. For a generating set with an output exceeding 16 A, the accessibility of this means of isolation shall comply with national rules and distribution system operator requirements. For a generating set with an output not exceeding 16 A, the accessibility of this means of isolation shall comply with BS EN 50438.*

15.5 Earthing

When a SSEG is operating in parallel with a distributor's network, there shall be no direct connection between the generator winding (or pole of the primary energy source in the case of PV array or fuel cell) and the network operator's earth terminal. See Figure 15.1.

▼ **Figure 15.1** Earthing of parallel operation SSEG

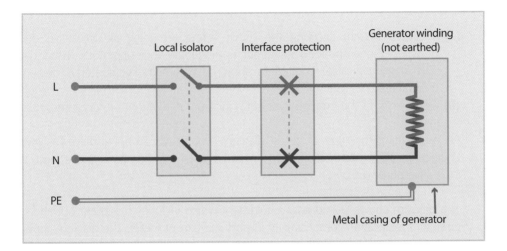

Chap 54 For all connections, earthing arrangements shall comply with the requirements of BS 7671.

15.6 Labelling

514.15.1 Labels (i.e. warning notices) are required at:

(a) the origin of the installation (fused cut-out);
(b) the meter position, if remote from the origin;
(c) the consumer unit or distribution board to which the SSEG is connected; and
(d) all points of isolation of all sources of supply, to indicate the presence of the SSEG within the premises.

The Health and Safety (Safety Signs and Signals) Regulations 1996 stipulate that the labels should display the prescribed triangular shape and size using black on yellow colouring. A typical label both for size and content is shown in Figure 15.2.

▼ **Figure 15.2** Warning notice for alternative supplies

In addition, Engineering Recommendation G83 requires up-to-date information to be displayed at the point of connection with a distributor's network as follows:

(a) a circuit diagram showing the circuit wiring, including all protective devices, between the embedded generator and the network operator's fused cut-out. This diagram is also required to show by whom the generator is owned and maintained.

(b) a summary of the protection settings incorporated within the equipment.

An example of the type of circuit diagram that needs to be displayed is given in Engineering Recommendation G83. The diagram is non-prescriptive and is for illustrative purposes only.

The diagram shows Distributor's equipment which includes fused cut out and meter operator's installation comprising an import and export meter. The diagram shows the consumers installation which includes the consumer unit, a lockable double-pole AC isolator, a feed in tariff meter, an AC double-pole isolator, an inverter, a DC double-pole isolator, and the SSEG with interface protection. Refer to G83 for more information.

The installer is required to advise that it is the user's responsibility to ensure that this safety information is kept up to date. The installation operating instructions shall contain the SSEG manufacturer's contact details, e.g. name, telephone number and web address.

Temporary electrical installations for structures, amusement devices and booths at fairgrounds, amusement parks and circuses

16

16.1 Introduction

Sect 740 There are no significant changes to Section 740 introduced by the 18th Edition.

Section 740 recognizes that some installations are exposed to many differing and onerous circumstances, as they are often frequently installed, dismantled, moved to a new location then installed and operated again.

To compound problems, such installations can be exposed to the elements, and the locations are open to the general public, may house animals and livestock, and are also a place of work.

The equipment must function without compromising safety; the installation has to, therefore, be fit for purpose and be designed to cope with ever-changing conditions.

Section 740 prescribes particular measures to reduce the risks caused by this harsh treatment of the electrical installation.

16.2 Scope

740.1.1 Section 740 specifies the minimum electrical installation requirements to facilitate the safe design, installation and operation of temporarily erected mobile or transportable electrical machines and structures that incorporate electrical equipment. The machines and structures are intended to be installed repeatedly, without loss of safety, temporarily, at fairgrounds, amusement parks, circuses or similar places.

Note: The permanent electrical installation, if there is one, is excluded from the scope.

The object of this section is to define the electrical installation requirements for such structures and machines, being either integral parts or constituting the total amusement device. The scope does not include the internal electrical wiring of machines (see BS EN 60204-1).

Note: Guidance on temporary electrical installations for entertainment and related purposes is given in BS 7909.

16.3 Electrical supplies

740.313.1.1 The nominal supply voltage of temporary electrical installations in booths, stands and amusement devices should not exceed 230/400 V AC or 440 V DC. Supplies can be obtained from a number of sources:

(a) from the public network, i.e. the DNO;
(b) generators, i.e. those mounted on trucks owned by the touring event; and
(c) from privately owned supplies, e.g. a local factory with sufficient spare capacity.

740.313.3 There can be any number of electrical sources supplying the temporary system and it is of paramount importance that line and neutral conductors from different sources are not interconnected.

Where the supply is obtained from the DNO any instructions given must be adhered to. Supplies obtained from the DNO would preferably be TN-S but this isn't always possible. A TN-S system has the neutral of the source of energy connected with Earth at one point only, which is at, or as near as is reasonably practicable to, the source of supply.

The consumer's main earthing terminal is typically connected to the metallic sheath of the distributor's SWA service cable.

740.411.4.1 A PME earthing facility shall not be used as the means of earthing for an installation falling within the scope of this section.

Note: The ESQCR prohibit the use of a PME earthing facility as the means of earthing for the installation of a caravan or similar construction.

740.411.4.3 Where the type of system earthing is TN, a PEN conductor shall not be used downstream of the origin of the temporary electrical installation.

740.411.6 Where continuity of service is important, IT systems may be used for DC applications only.

16.4 Protection against electric shock

740.410.3 At the origin of each electrical supply, to all or part of the installation, an RCD, with a rated residual operating current not exceeding 300 mA, is to be installed to provide automatic disconnection of supply. As there will be further RCDs downstream of this point, this RCD should be of the S-type complying with the requirements of BS EN 61008-1 or BS EN 61009-1, and incorporate a time delay in accordance with BS EN 60947-2, to provide selectivity with the RCDs protecting final circuits.

740.411 For supplies to AC motors, RCDs should be of the time-delayed or S-type where necessary to prevent unwanted tripping.

740.410.3.5 The protective measure of obstacles (Regulation 417.2) shall not be used in this type of installation; however, placing out of arm's reach is acceptable for electric dodgems (see section 16.8.2).

740.410.3.6 The protective measures of non-conducting location (Regulation 418.1) and earth-free local equipotential bonding (Regulation 418.2) shall not be used.

16.5 Additional protection

740.415.1 All final circuits in the installation for lighting, socket-outlets rated up to 32 A, and mobile equipment connected by a flexible cable and rated up to 32 A, are to be protected by RCDs having a rated residual operating current not exceeding 30 mA.

The requirement for additional protection relates to the increased risk of damage to cables within an installation of this nature.

Lighting circuits incorporating emergency luminaires, with, for example, self-contained batteries should be protected by the same RCD protecting that lighting circuit.

Additional protection is not required for:

(a) SELV or PELV circuits – this measure alone is deemed to be a protective measure in all situations;
(b) circuits protected by electrical separation; or
(c) lighting circuits placed out of arm's reach – provided they are not supplied by socket-outlets for household or similar purposes, e.g. those manufactured to BS 1363, or socket-outlets according to BS EN 60309-1. Typically, this would be achieved by devices for connection of luminaires to BS EN 61995 or luminaire supporting couplers to BS 6972.

740.415.2 To minimize potentials, supplementary protective equipotential bonding must be installed to connect all exposed-conductive-parts and extraneous-conductive-parts that can be touched by livestock. Where a metal grid is laid in the floor, or extraneous-conductive-parts are accessible, they should be included within the supplementary bonding of the location. It is important to note that animal excrement and urine are very corrosive and so all supplementary bonding connections should be protected by suitable enclosure.

16.6 Installation

16.6.1 Wiring systems

740.521.1 Conduit, cable trunking, cable ducting, tray and ladder systems should comply with the following standards:

(a) conduit systems – BS EN 61386 series;
(b) cable trunking and ducting systems – the relevant part 2 of BS EN 50085; and
(c) tray and ladder systems – BS EN 61537.

Buried cables should be protected against mechanical damage. Conduit classified as 450N for protection against compression and classified as normal for protection against impact, according to BS EN 61386-24, would fulfil this requirement.

Mechanical protection should be used in all public areas and in areas where the wiring system crosses roads or walkways. Where mechanical protection is provided:

(a) conduit systems should comply with BS EN 61386-21 with a classification of heavy protection against compression and heavy protection against impact. Metallic and composite conduit systems should be class 3 for protection against corrosion, i.e. medium protection inside and high protection outside.
(b) cable trunking and cable ducting systems should comply with the BS EN 50085 series with a classification 5J for protection against impact.

Where subject to movement, the wiring system should be of flexible construction and should comply with BS EN 61386-23. This should use cables of type H07RN-F or H07BN4-F (BS EN 50525-2-21).

16.6.2 Cables

740.521.1 All cables should be fire rated and meet the requirements of BS EN 60332-1-2. Cables of type H07RN-F or H07BN4-F (BS EN 50525-2-21) together with conduit complying with BS EN 61386-23 are deemed to satisfy this requirement.

Cables should have a minimum rated voltage of 450/750 V, except that, within amusement devices, cables having a minimum rated voltage of 300/500 V may be used.

Where cables are buried in the ground, the route should be marked at suitable intervals and be protected against mechanical damage. Where there is a risk of mechanical damage, armoured cables or cables protected against mechanical damage due to external influence of medium severity or greater (e.g. > AG2) should be used.

16.6.3 Electrical connections

740.526 Joints must not be made in cables except where used as a necessary connection into a circuit. Where joints are made, these should either use connectors in accordance with the relevant British Standards and the manufacturer's instructions or the connection should be made in an enclosure with a degree of protection of at least IPXXD or IP4X rating. Where strain can be transmitted to terminals the connection should incorporate cable anchorage(s).

16.6.4 External influences

740.512.2 Electrical equipment should have a degree of protection of at least IP44 rating.

16.6.5 Switchgear and controlgear

740.51
Part 2
Switchgear and controlgear should be placed in cabinets that can be opened only by the use of a key or a tool, except for those parts designed and intended to be operated by ordinary persons.

16.6.6 Isolation, switching and overcurrent protection

740.537.1 It is a requirement that every electrical installation of a booth, stand or amusement device has its own means of isolation, switching and overcurrent protection; these devices should be readily accessible.

740.537.2.1.1 There are similar isolation requirements for distribution circuits supplying outdoor installations.

740.537.2.2 A device for isolation must disconnect all live conductors – line(s) and neutral conductors. Examples of devices that can be used for isolation where marked as suitable are:

(a) circuit-breakers;
(b) RCDs; and
(c) plug and socket arrangements.

Table 537.4 See Table 537.4 of BS 7671:2018, which includes guidance on device types that are suitable for isolation.

16.6.7 Luminaires

740.55.1.1 Every luminaire and decorative lighting chain should have a suitable IP rating and be securely attached to the structure or support intended to carry it. Its weight should not be carried by the supply cable, unless it has been selected and erected for this purpose. Luminaires and decorative lighting chains mounted less than 2.5 m above floor level, i.e. within arm's reach, or which could be otherwise accessible to accidental contact, should be firmly fixed, sited and guarded to prevent risk of injury to persons or ignition of materials.

Access to the fixed light source should only be possible after removing a barrier or an enclosure, which should only be possible by the use of a tool.

Lighting chains should use H05RN-F or H07RN-F (BS EN 50525-2-21) cable or equivalent, and may be used in any length provided the overcurrent protective device in the circuit is correctly rated.

740.55.1.2 Insulation-piercing lampholders must not be used unless the cables and lampholders are compatible and the lampholders are non-removable once fitted to the cable.

740.55.3 Luminous tubes, signs or lamps with an operating voltage higher than 230 V/400 V AC, e.g. neon signs, are to be installed out of arm's reach or be adequately protected from accidental or deliberate damage, to reduce the risk of injury to persons.

A separate circuit must be used, which must be controlled by an emergency switch.

The switch should be easily visible, accessible and marked in accordance with the requirements of the local authority.

740.55.1.3 Lamps in shooting galleries and other sideshows where projectiles are used must be suitably protected against accidental damage.

740.55.1.4 Where transportable floodlights are used, they must be mounted so that the luminaire is inaccessible to non-instructed persons. Supply cables must be flexible and have adequate protection against mechanical damage.

16.6.8 Safety isolating transformers and electronic convertors

740.55.5 Safety isolating transformers must comply with BS EN 61558-2-6 or provide an equivalent degree of safety.

Each transformer or electronic convertor must incorporate a protective device that can only be manually reset; this device should protect the secondary circuit.

Safety isolating transformers should be mounted out of arm's reach or be mounted in a location that provides equal protection, e.g. in a panel or room (with adequate ventilation) that can only be accessed by skilled or instructed persons, to facilitate inspection, testing and maintenance.

Electronic convertors should conform to BS EN 61347-2-2.

Enclosures containing rectifiers and transformers must be adequately ventilated and the vents must not be obstructed when in use.

16.6.9 Plugs and socket-outlets

740.55.7 An adequate number of socket-outlets must be installed to allow user requirements to be met safely.

In booths, stands and for fixed installations, one socket-outlet for each square metre or linear metre of wall is generally considered adequate.

Socket-outlets dedicated to lighting circuits placed out of arm's reach should be labelled according to their purpose.

When used outdoors, plugs, socket-outlets and couplers must comply with BS EN 60309-2, or, where interchangeability is not required, BS EN 60309-1.

16.7   Fire risk

GN4 Additional guidance for protection against fire for this type of location can be found in section 8.4 of Guidance Note 4: *Protection Against Fire*.

16.7.1 Luminaires and floodlights

740.55.1.5 Luminaires and floodlights must be so fixed and protected that a focusing or concentration of heat is not likely to cause ignition of any material.

16.7.2 Electric motors

422.3.7 An electric motor that is automatically or remotely controlled or which is not continuously supervised must be fitted with a manual reset protective device against excess temperature.

16.8 Electrical equipment

16.8.1 Electrical supply to devices

740.55.8 At each amusement device, there must be a connection point readily accessible and permanently marked to indicate the following essential characteristics:

(a) rated voltage;
(b) rated current; and
(c) rated frequency.

16.8.2 Electric dodgems

740.55.9 Electric dodgems must only be operated at voltages not exceeding 50 V AC or 120 V DC. The circuit should have electrical separation from the supply by means of an isolating transformer in accordance with BS EN 61558-2-4 or a motor-generator set.

16.8.3 Low voltage generating sets

740.551 It is very important that all generators are located or protected so as to prevent danger and injury to people through inadvertent contact with hot surfaces and dangerous parts.

The electrical equipment associated with the generator must be mounted securely and, if necessary, on anti-vibration mountings.

542.2.4 Where a generator supplies a temporary installation, forming part of a TN, TT or IT system, care should be taken to make certain that the earthing arrangements are adequate and, in cases where earth electrodes are used, that they are considered to be continuously effective. In reality, this means that the drying of the ground in summer, or freezing of the ground in winter, should not adversely affect the value of earth fault loop impedance for the installation. The neutral conductor of the star-point of the generator should, except for an IT system, be connected to the exposed-conductive-parts of the generator.

16.9 Inspection and testing

16.9.1 The temporary installation

740.6 The electrical installation between its origin and any electrical equipment must be inspected and tested after each assembly on site.

Internal electrical wiring of roller coasters, electric dodgems, etc., is not considered as part of the verification of the electrical installation. In special cases the number of the tests may be modified according to the type of temporary electrical installation.

16.9.2 Amusement devices

The scope of Section 740 does not cover amusement devices but, in law, there is still a requirement to ensure that the devices are fit for use.

The Amusement Device Safety Council (ADSC) is the policy-making body for safety, self-regulation and technical guidance in the UK amusement industry. This committee, in partnership with HSE, develops the policy for and oversees the Amusement Device Inspection Procedures Scheme (ADIPS). ADIPS is the fairground and amusement park industry's self-regulated safety inspection scheme, which registers competent ride inspectors and the rides they inspect. The purpose of the scheme is to promote and improve fairground and amusement park safety through rules and procedures relating to the annual inspection of the amusement devices. It is supported by industry associations who require that their members use ADIPS. ADIPS is also available to any operator of amusement devices as it demonstrates 'best practice' and their compliance with the Health and Safety at Work etc. Act 1974.

Operating and maintenance gangways

17

Sect 729 ## 17.1 Introduction

There are no significant changes to Section 729 introduced by the 18th Edition.

Sect 729
132.12 Section 729 applies to restricted areas. These are areas such as switchrooms with switchgear and controlgear assemblies with a need for operating or maintenance gangways for authorized persons. The fundamental requirement for accessibility of electrical equipment in BS 7671:2018 is contained in Chapter 13. Regulation 132.12 states:

Electrical equipment shall be arranged to provide:

(i) sufficient space for the initial installation and later replacement of individual items of electrical equipment
(ii) accessibility for operation, inspection, testing, fault detection, maintenance and repair.

Regulation 15 of the Electricity at Work Regulations 1989 has requirements for working space, access and lighting and requires that, for the purposes of enabling injury to be prevented, adequate working space, adequate means of access and adequate lighting shall be provided at all electrical equipment on which or near which work is being done in circumstances that may give rise to danger (Regulation 15 applies whilst there is work activity ongoing). Regulation 14 of the Electricity at Work Regulations is concerned with work on or near any live conductors.

Sect 417 See also Section 417 of BS 7671:2018 about the protective measures of obstacles and placing out of reach.

Section 729 prescribes particular measures to reduce the risks in these restricted areas.

17.2 Scope

729.1 The particular requirements of this section apply to basic protection and other aspects relating to the operation or maintenance of switchgear and controlgear within areas including gangways, where access is restricted to skilled or instructed persons.

729.513 **17.3 Accessibility**

17.3.1 Requirements for operating and maintenance gangways

729.513.2 Regulation 729.513.2 requires that the width of gangways and access areas shall be adequate for work, operational access, emergency access, emergency evacuation and for transport of equipment. Gangways shall permit at least a 90 degree opening of equipment doors or hinged panels.

17.3.2 Restricted access areas where basic protection is provided by barriers or enclosures

729.513.2.1 In restricted access areas where basic protection is provided by barriers or enclosures, Regulation 729.513.2.1 gives the following minimum dimensions (see Fig 17.1):

(a) Gangway width of 700 mm between: barriers or enclosures and switch handles or circuit-breakers in the most onerous position, and barriers or enclosures or switch handles or circuit-breakers in the most onerous position and the wall

(b) Gangway width of 700 mm between barriers or enclosures or other barriers or enclosures and the wall

(c) Height of gangway to barrier or enclosure above floor, minimum dimension 2000 mm

417.3 **(d)** Live parts placed out of reach, see Regulation 417.3 (minimum dimension 2500 mm).

Note: Where additional workspace is needed e.g. for special switchgear and controlgear assemblies, larger dimensions may be required.

17.3.3 Restricted access areas where the protective measure of obstacles is applied

729.513.2.2 In restricted access areas where the protective measure of obstacles applies, Regulation 729.513.2.2 gives the following minimum dimensions (see Fig 17.2):

(a) Gangway width of 700 mm between: obstacles and switch handles or circuit-breakers in the most onerous position, and obstacles or switch handles or circuit-breakers in the most onerous position and the wall

(b) Gangway width of 700 mm between obstacles or other obstacles and the wall

(c) Height of gangway to obstacles above floor, minimum dimension 2000 mm

417.3 **(d)** Live parts placed out of reach, see Regulation 417.3 (minimum dimension 2500 mm).

Note: The HSE would not expect to see any new switchboards installed in the UK where there was potential access to live exposed busbars.

17.3.4 Access to gangways

729.513.2.3 Regulation 729.513.2.3 has requirements for access to gangways. For closed restricted access areas with a length exceeding 6 m, accessibility from both ends is recommended. However, gangways longer than 10 m must be accessible from both ends.

The regulation recognizes that this may be accomplished by placement of the equipment a minimum of 700 mm from all walls or by providing an access door, if needed, on the wall against which the equipment is positioned. However, closed restricted access areas with a length exceeding 20 m must be accessible by doors from both ends.

Annex A729 Finally, Annex A729 (note: normative, i.e. obligatory) contains a number of additional requirements for closed restricted access areas in order to permit easy evacuation. The annex considers two cases; these require:

▶ a minimum passing width of 700 mm with circuit-breakers in the isolated position and all doors closed; and

▶ a minimum passing width of 500 mm with circuit-breakers fully withdrawn and doors fully open.

Persons involved in this work are advised to seek advice from the HSE.

▼ **Figure 17.1** (Fig 729.1 of BS 7671) Gangways in installations with protection by barriers or enclosures

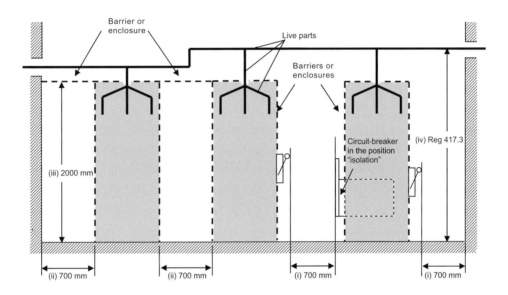

Note: The above dimensions apply after barriers and enclosures have been fixed and with circuit-breakers and switch handles in the most onerous position, including "isolation".

▼ **Figure 17.2** (Fig 729.2 of BS 7671) Gangways in installations with protection by obstacles

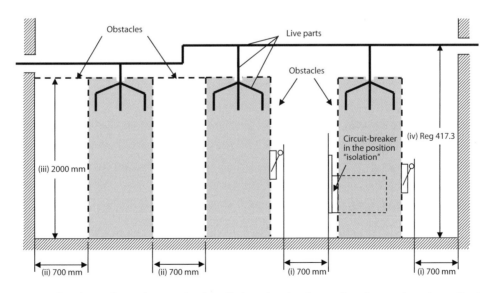

Note: The above dimensions apply after all obstacles, barriers and enclosures have been fixed and with circuit-breakers and switch handles in the most onerous position, including "isolation".

Outdoor lighting installations

18

18.1 Introduction

Sect 714 There are no significant changes to Section 714 introduced by the 18th Edition.

18.2 Scope

714.1 Outdoor lighting installations include those for roads, parks, car parks, gardens, places open to the public, sporting areas, illumination of monuments and floodlighting, as well as other lighting arrangements in places such as telephone kiosks, bus shelters, advertising panels and town plans. Road signs are also included, but not road traffic signal systems.

Section 714 does not include temporary festoon lighting. Furthermore, luminaires fixed to the outside of a building and supplied directly from the internal wiring of that building are not included as these are considered part of the building electrical installation.

18.3 Protection against electric shock

714.410.3.6 As would be expected, the protective measures of a non-conducting location (Regulation 418.1) and earth-free local equipotential bonding (Regulation 418.2) shall not be used. These measures are contained in Section 418 of BS 7671:2018 and are not for general application, but instead they are for application only in installations controlled or supervised by skilled or instructed persons. Equipment used for highway power supplies is usually located in areas accessible to the public.

714.411.2.201 **Provisions for basic protection – equipment doors and luminaires**

Doors providing access to electrical equipment contained in street furniture provide a measure of protection against interference; however, the likelihood of removal or breakage is such that a door less than 2.5 m above ground level cannot be relied upon to provide basic protection. Equipment or barriers within the street furniture are required to prevent contact with live parts by a finger (IP2X or IPXXB rating).

For a luminaire at a height of less than 2.80 m above ground level, access to the light source shall only be possible after removing a barrier or an enclosure requiring the use of a tool.

714.411.202 **Highway power supply circuits**

Regulation 714.411.202 modifies the general requirements, and permits a maximum disconnection time of 5 s for all circuits feeding fixed equipment used in highway power supplies in line with present day practice for compliance with Regulation 411.3.2.3 (TN system) or 411.3.2.4 (TT system).

714.411.203 **PME**

Regulation 714.411.203 requires that where an earth connection to a distributor's PME network has been provided for a street electrical fixture, the earthing and bonding conductor of a street electrical fixture shall have a minimum copper equivalent cross-sectional area of 6 mm^2 for supply neutral conductors up to 10 mm^2, or in accordance with Table 54.8 for larger sizes. The requirement of 6 mm^2 is based on the Energy Networks Association Engineering Recommendation G12/4 (requirements for the application of protective multiple earthing (PME) to low voltage networks).

18.4 Operational conditions and external influences

714.512.2 Any wiring system or equipment selected and installed must be suitable for its location and able to operate satisfactorily without deterioration during its working life. The presence of water can occur in several ways, e.g. rain, splashing, steam/humidity, condensation, and at each location where it is expected to be present its effects must be considered. Suitable protection must be provided, both during construction and for the completed installation. Regulation 714.512.2.105 requires that electrical equipment shall have, by construction or by installation, a degree of protection of at least IP33 rating in order to protect against sprays (AD3). The IP classification code, BS EN 60529:2004, describes a system for classifying the degrees of protection provided by the enclosures of electrical equipment. The degree of protection provided by an enclosure is indicated by two numerals. The first numeral indicates protection of persons against access to hazardous parts inside enclosures or protection of equipment against ingress of solid foreign objects. The second numeral indicates protection of equipment against ingress of water. More information on the IP classification code is given in IET Guidance Note 1: *Selection & Erection*.

Notices

714.514.12 Regulation 714.514.12.201 contains a relaxation for notices. The general requirements for a periodic inspection and testing notice and an RCD testing notice need not be applied where the installation is subject to a programmed inspection and testing procedure. However, there is an additional requirement for temporary supplies. Regulation 714.514.12.202 requires an externally mounted durable label on every temporary supply unit stating the maximum sustained current to be supplied from that unit.

18.5 Isolation and switching

714.537 Each item of street furniture is required to have a local individual means of isolation. The established practice of using the fuse carrier as the isolation and switching device is allowed for TN systems provided only instructed persons carry out the work. Formal instruction with the issue of authorization is appropriate, particularly if the electricity distributor's consent is required (see Regulation 714.537.2.1.202). Reference in Regulation 714.537.2.1.201 to a suitably rated fuse carrier will generally limit the use of the fuse carrier as the main switch to individual items of equipment, e.g. one street light, as the fuse carrier should generally not be used to switch currents exceeding 16 A.

18.6 Bus shelters, etc. and decorative lighting

714.411.3.3 Lighting in bus shelters, telephone kiosks and similar must be provided with 30 mA RCD protection.

It is recommended that any decorative lighting (such as Christmas tree lights) within reach of the general public be supplied by SELV. Such equipment is readily available and suitable for indoor and outdoor use.

18.7 Protection and identification of cables

The definitions in Regulation 1(5) of the ESQCR are intended to include cables supplying street furniture within the scope of the statutory instrument. Such cables are included within the term 'network' and the persons owning or operating such cables are 'distributors' as defined. The implications of this are as follows:

▶ in accordance with Regulation 14, the cables must be buried at sufficient depth to prevent danger and they must be protected or marked; and

▶ in accordance with Regulation 15(2), the distributor must maintain maps of the cables.

The Department for Business, Energy and Industrial Strategy (formerly the DTI) guidance on the ESQCR gives advice on the methods that dutyholders should employ to demonstrate compliance with Regulation 14(3) and to thereby reduce the risk of injury to contractors or members of the public. Listed in order of preference the methods are:

(a) cable installed in a duct with marker tape above;

(b) cable installed in a duct only;

(c) cable laid direct and covered with protective tiles;

(d) cable laid direct and covered with marker tape; and

(e) some other method of mark or indication.

In consideration of the methods by which cables should be marked or protected, dutyholders should make allowance for the environment in which the cables are installed and the risks to those who may need to expose and work on or near the cables in future. The National Joint Utilities Group (NJUG) have agreed colours for ducts, pipes, cables and marker/warning tapes when laid in the public highway, see Tables 5.2 and 5.3 in Chapter 5 of IET Guidance Note 1: *Selection & Erection*.

18.8 References

▶ BS 5489 *Code of practice for the design of road lighting. Lighting of roads and public amenity areas*.

▶ NJUG Publication 7 Guidelines on the positioning and colour coding of underground utilities' apparatus.

▶ Energy Networks Association Engineering Recommendation G39 issue 2 2013, Model code of practice covering electrical safety in the planning, installation, commissioning and maintenance of public lighting and other street furniture.

▶ Institution of Lighting Engineers *Code of practice for electrical safety in public lighting operations*, 4th edition.

Extra-low voltage lighting installations

19

19.1 Introduction

Section 715 There are no significant changes to Section 715 introduced by the 18th Edition.

19.2 Scope

715.1 The particular requirements of this section apply to extra-low voltage lighting installations supplied from a source with a maximum rated voltage of 50 V AC or 120 V DC.

19.3 Protection against electric shock

715.414 Regulation 715.414 requires that an extra-low voltage luminaire without provision for the connection of a protective conductor shall be installed only as part of a SELV system. The protective measure of FELV shall not be used (Regulation 715.411.7.201).

Where bare conductors are used, the nominal voltage shall not exceed 25 V AC or 60 V DC according to Regulation 414.4.5. Where the SELV source of a SELV lighting system is a safety isolating transformer, it should meet the requirements of BS EN 61558-2-6 and at least one of the requirements of Regulation 715.422.106. Regulation 715.414 only permits parallel operation of transformers in the secondary circuit if they are also paralleled in the primary circuit and the transformers have identical electrical characteristics.

Regulation 715.414 requires that an electronic convertor used for an extra-low voltage lighting installation must comply with BS EN 61347-2-2 Annex 1 for incandescent lamps or BS EN 61347-2-13 Annex 1 for LEDs. The regulation prohibits parallel operation of convertors to BS EN 61347-2-2 or BS EN 61347-2-13.

19.4 Protection against thermal effects – fire risk of transformers/convertors

715.42 Transformers shall be either:

(a) protected on the primary side by a protective device complying with the requirements of 715.422.107.2; or

(b) short-circuit proof (both inherently and non-inherently).

Fire risk from short-circuiting of uninsulated conductors

If both circuit conductors are uninsulated, they must be either:

(a) provided with a protective device complying with all the requirements (a) to (e) below; or

(b) supplied from a transformer or convertor, the power of which does not exceed 200 VA; or

(c) the system must comply with BS EN 60598-2-23.

Where the first option is adopted to provide protection against the risk of fire, the device must:

(a) continuously monitor the power demand of the luminaires;

(b) automatically disconnect the supply circuit within 0.3 s in the event of a short-circuit or failure that causes a power increase of more than 60 W;

(c) automatically disconnect while the supply circuit is operating with reduced power (for example, by gating control or a regulating process or a lamp failure) if there is a failure that causes a power increase of more than 60 W;

(d) automatically disconnect during switching of the supply circuit if there is a failure that causes a power increase of more than 60 W;

(e) be fail-safe.

19.5 Protection against overcurrent

715.430.104 SELV circuits must be protected against overcurrent either by a common protective device or by a protective device for each SELV circuit, in accordance with the requirements of Chapter 43. The use of self-resetting overcurrent protective devices is permitted only for transformers up to 50 VA.

19.6 Types of wiring system

715.521.1 The following wiring systems should be used:

(a) insulated conductors in conduit to the BS EN 61386 series or trunking or ducting to the BS EN 50085 series;

(b) rigid cables;

(c) flexible cables;

(d) systems for extra-low voltage lighting according to BS EN 60598-2-23;

(e) track systems according to BS EN 60570; and

(f) bare conductors (see Regulation 715.521.106).

Where parts of the extra-low voltage lighting installation are accessible, the requirements of Section 423 of BS 7671 also apply. Metallic structural parts of buildings, e.g. pipe systems or parts of furniture, must not be used as live conductors.

19.7 Bare conductors

715.521.106 If the nominal voltage does not exceed 25 V AC or 60 V DC, bare conductors may be used provided the extra-low voltage lighting installation complies with all the following requirements:

(a) the lighting installation is designed and installed or enclosed in such a way that the risk of a short-circuit is reduced to a minimum;

(b) the conductors used have a cross-sectional area in accordance with Regulation 715.524; and

(c) the conductors or wires are not placed directly on combustible material.

For suspended bare conductors, at least one conductor and its terminals must be insulated for that part of the circuit between the transformer and the protective device to prevent a short-circuit.

19.8 Suspended systems

715.521.107 | If the fixing accessory is intended to support a pendant luminaire, the accessory should be capable of carrying a mass of not less than 5 kg. If the mass of the luminaire is greater than 5 kg, the installer should ensure that the fixing means is capable of safely supporting its mass. The installation instructions of the manufacturer should be followed. Termination and connection of conductors should be made by screw terminals or screwless clamping devices complying with BS EN 60998-2-1 or BS EN 60998-2-2. Insulation piercing connectors and termination wires, with counterweights, hung over suspended conductors, should not be used. The suspended system should be fixed to walls or ceilings by insulated distance cleats and be continuously accessible throughout the route.

19.9 Cross-sectional area of conductors

715.524.201 | The minimum cross-sectional area of the ELV conductors for connection to the output terminals or terminations of transformers/convertors shall be chosen according to the load current.

For systems with luminaires suspended from the conductors, the minimum cross-sectional area of the ELV conductors for connection to the output terminals or terminations of transformers/convertors shall be 4 mm², for mechanical reasons.

19.10 Voltage drop in consumers' installations

715.525 | Regulation 715.525 modifies the general rules relating to voltage drop contained in Section 525 and Table 4Ab of Appendix 4 which gives a voltage drop of 3 % for lighting.

Regulation 715.525 allows a voltage drop between the transformer and the furthest luminaire of 5 % (of the nominal voltage of the ELV installation) for compliance with Section 525.

19.11 Protective devices and SELV sources

715.530.3.104 | Regulation 715.530.3.104 recognizes that protective devices may be located above false ceilings, which are removable or easily accessible, provided that information is given making people aware of their location and the circuits that the protective devices control. SELV sources and protective devices shall be permanently connected. The regulation reinforces the general rules for adequate support, the need to avoid overheating due to thermal insulation and the need to avoid mechanical stress on electrical connections.

19.12 Isolation and switching

715.537.1.1 Where transformers are operated in parallel, the primary circuits must be permanently connected to a common isolating device (see Figure 19.1). The transformers must also have identical electrical characteristics.

▼ **Figure 19.1** SELV lighting installation with parallel-connected transformers and a common output circuit

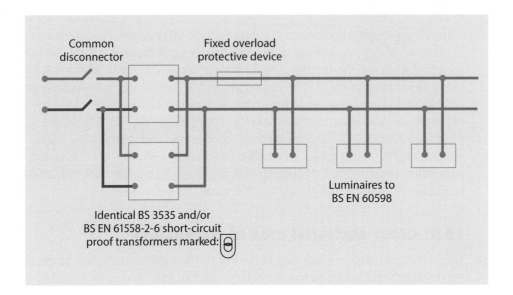

Electric vehicle charging installations

20

20.1 Introduction

Sect 722 The 18th Edition introduces a number of significant changes in Section 722. These include changes to Regulation 722.411.4.1 concerning the use of a PME supply, for which the exception concerning dwellings has been deleted. Changes have also been made to requirements for external influences, RCDs, socket-outlets and connectors.

20.2 Scope

Section 722 provides requirements for circuits intended to supply electric vehicles (EVs) for charging purposes. Inductive charging and charging of mobility scooters and similar vehicles of 10 A and less are not covered.

Protection for safety when feeding back electricity from an electric vehicle into a private or public network is under consideration. The requirements are still being developed at European level.

Note: A fully revised IET *Code of Practice for Electric Vehicle Charging Equipment Installation, 3rd Edition* is now available

20.3 Protection against electric shock

As would be expected, the protective measures of obstacles, placing out of reach, non-conducting location and earth-free local equipotential bonding shall not be used. These measures are contained in Sections 417 and 418 of BS 7671:2018 and are not for general application. The protective measures of Section 417 provide basic protection only and are for application in installations controlled or supervised by skilled persons. The fault protective provisions of Section 418 are special and, again, subject to control and effective supervision by skilled or instructed persons.

722.413.1.2 The protective measure of electrical separation is permitted, but must be limited to the supply of one electric vehicle. The circuit must be supplied through a fixed isolating transformer complying with BS EN 61558-2-4. IET Guidance Note 5: *Protection Against Electric Shock* gives details on electrical separation.

20.4 Protective multiple earthing

The ESQCR 2002 permit the distributor to combine neutral and protective functions in a single conductor provided that, in addition to the neutral to Earth connection at the supply transformer, there are one or more other connections with Earth. The supply neutral may then be used to connect circuit protective conductors of the customer's installation with Earth, if the customer's installation meets the requirements of BS 7671. This protective multiple earthing (PME) has been almost universally adopted by distributors in the UK as an effective and reliable method of providing their customers with an earth connection. However, under certain supply system fault conditions (external to the installation) a potential can develop between the conductive parts connected to the PME earth terminal and the general mass of Earth. The potential difference between true Earth and the PME earth terminal is of importance when:

(a) body contact resistance is low (little clothing is worn, conditions are damp or wet); and/or

(b) there is relatively good contact with true Earth.

Contact with Earth is always possible outside a building. If exposed-conductive-parts and/or extraneous-conductive-parts connected to the PME earth terminal are accessible outside the building, people may be subjected to a voltage difference appearing between these parts and Earth.

722.411.4.1 Regulation 722.411.4.1 does not allow PME as a means of earthing for an electric vehicle charging point where the charging point or the vehicle is located outdoors, except where Regulation 722.411.4.1(i), (ii) or (iii) apply.

> **Note:** A number of options are explained in the IET *Code of Practice for Electric Vehicle Charging Equipment Installation.*

A single-phase installation: Regulation 722.411.4.1(ii)

For Regulation 722.411.4.1(ii) to be met and apply, and therefore allow connection to a PME system, a very low earth electrode resistance must be achieved in order to mitigate the effects of a potential open-circuit PEN conductor fault on the supply.

A722.3 For the purposes of condition (ii) of Regulation 722.411.4.1, the sum of the resistances of the earth electrode and the protective conductor connecting it to the main earthing terminal must meet the following condition, as applicable.

For a single-phase installation:

$$R_{A\,ev} \le \frac{70\,U_0}{I_{inst}\,(U_0 - 70)}$$

where:

$R_{A\,ev}$ is the sum of the resistances of the earth electrode and the protective conductor connecting it to the main earthing terminal of the installation (in ohms).

U_0 is the nominal AC rms line voltage to Earth.

I_{inst} is the rms maximum demand current of a single-phase installation (in amps), including that of the electric vehicle charging load and any other loads, determined in accordance with Regulation 311.1.

Example:

For example purposes, if we take a resistive load of a dwelling as 2 kW (8.7 A), and the maximum charging current of an electric vehicle as 10 A, then we may assume a total installation load (I_{inst}) of 18.7 A.

Therefore, for a single-phase installation:

$$R_{A\,ev} \le \frac{70 \times 230}{18.7 \times (230 - 70)} = 5.4\ \Omega$$

Where it is required to predict the resistance to earth of a vertical rod, the following equation can be used:

$$R = \frac{\rho}{2\pi L}\left[\log_e\left(\frac{8L}{d}\right) - 1\right](\Omega)$$

where:

L = the length of the electrode in metres
d = the diameter of the electrode in metres
ρ = the resistivity of the soil in ohm metres (Ωm)

Example:

L = 1.8 m long rod
d = 15 mm diameter rod
ρ = 10 for clay type soil from BS 7430

Therefore, calculating the resistance

$$R = \frac{10}{2\pi \times 1.8} \times \left[\ln\left(\frac{8 \times 1.8}{0.015}\right) - 1\right] = 5.18\ \Omega$$

The above equation predicts that a 15 mm diameter rod electrode buried to a depth of 1.8 m may not produce a low enough resistance (see Note 2 on the following page). The diameter of the rod or its length can be changed so as to reduce the predicted resistance value.

In this further example we use a rod of twice the buried length of the original one (i.e. 3.6 m).

$$R = \frac{10}{2\pi \times 3.6} \times \left[\ln\left(\frac{8 \times 3.6}{0.015}\right) - 1\right] = 2.899\,\Omega$$

Note 1: Two separate rods in parallel and sufficiently spaced may have produced a better result.

Note 2: This is just the resistance of the earth electrode. We must add to this figure the resistance of the protective conductor connecting it to the main earthing terminal to check if we meet condition (ii) of Regulation 722.411.4.1.

Important: The values used are for example purposes only. It is the responsibility of the designer to determine I_{inst} and the sum of the resistances of the earth electrode and the protective conductor connecting it to the main earthing terminal.

For example, if we use a larger value of I_{inst}, such as 80 A, then using the same formula we get a value of $R_{Aev} = 1.257\ \Omega$. This will require a much lower earth electrode resistance, which may not be possible.

The Energy Networks Association may be able to help with determining I_{inst}.

Alternatives to using a PME supply are discussed in the IET *Code of Practice for Electric Vehicle Charging Equipment Installation*.

Where a charging point forms part of a three-phase installation

Regulation 722.411.4.1(i) states:

The charging point forms part of a three-phase installation that also supplies loads other than for electric vehicle charging and, because of the characteristics of the load of the installation, the maximum voltage between the main earthing terminal of the installation and Earth in the event of an open-circuit fault in the PEN conductor of the low voltage network supplying the installation does not exceed 70 V rms.

A722.2   Where triple harmonics can be neglected, condition (i) of Regulation 722.411.4.1 may be assumed to apply where the following condition is met:

$$\frac{I_m \times U_0}{I_{L1} + I_{L2} + I_{L3}} \leq 70$$

(In the above expression, I_m is the rms maximum neutral current of a three-phase installation (in amps), including that of the electric vehicle charging load and any other loads, determined in accordance with Regulation 311.1.)

If balanced, 50 Hz 3-phase load currents cancel out in the neutral, because of the 120-degree displacement of each phase. Even if the load is not balanced, the neutral will only carry the out-of-balance current.

Note: Regulation 722.411.4.1(i) will probably be met (and therefore apply) if the phases of the three-phase installation are balanced (equally loaded). Therefore, where (i) applies, the regulation permits a PME facility to be used.

If we put some typical figures into the formula, we see that even in a slightly out-of-balance situation, condition (i) applies and PME is permitted:

If we consider an unbalanced system with values of:

$I_{L1} = 80$ A
$I_{L2} = 100$ A
$I_{L3} = 65$ A
$N = 30.41$ A

$$\frac{30.41 \times 230}{80 + 100 + 65} = 28.55\ V$$

The result is less than 70, therefore PME is permitted.

Where protection is provided by a voltage-operated device to meet Regulation 722.411.4.1(iii)

Note: At the time of publication of this Guidance Note, it is understood that this device is not commercially available.

Regulation 722.411.4.1(iii) also requires an earth electrode (but not as low a resistance as condition (ii) of Regulation 722.411.4.1). Regulation 722.411.4.1(iii) is based on the installation requirements for a voltage-operated device. An important change in the 18th Edition is that the regulation now makes the point that this device could be included within the charging equipment and therefore may be provided by charging equipment manufacturers.

The touch voltage threshold of 70 V rms mentioned in Regulation 722.411.4.1(i), (ii) and (iii) is on the basis that Table 2c (Ventricular fibrillation for alternating current 50/60 Hz) of IEC 60479-5{ed1.0} gives a value of 71 V for both-hands-to-feet in water-wet conditions with medium contact area (12.5 cm^2).

20.5 Operational conditions and external influences

722.512 Any wiring system or equipment selected and installed must be suitable for its location and able to operate satisfactorily during its working life. The presence of water can occur in several ways, for example, from rain, splashing, steam/humidity or condensation. At each location where it is expected to be present, its effects must be considered. Suitable protection must be provided, both during construction and for the completed installation.

Regulation group 722.512.2 has been expanded in BS 7671:2018. Requirements are included for presence of water (AD), presence of solid foreign bodies (AE) and impact (AG).

Regulation 722.512.2.201 requires that where installed outdoors, the equipment shall be selected with a degree of protection of at least IPX4 in accordance with BS EN 60529 to protect against water splashes (AD4).

Regulation 722.512.2.202 requires that where installed outdoors, the equipment shall be selected with a degree of protection of at least IP4X in accordance with BS EN 60529 to protect against the ingress of very small objects (AE3).

Regulation 722.512.2.203 requires that where equipment is installed in public areas and car park sites, it shall be protected against mechanical damage (impact of medium severity AG2) by one or more of the following: location and/or mechanical protection and/or compliance with minimum degree of protection against mechanical impact of IK07 in accordance with the requirements of BS EN 62262.

The IP classification code BS EN 60529:2004 describes a system for classifying the degrees of protection provided by the enclosures of electrical equipment. The degree of protection provided by an enclosure is indicated by two numerals. The first numeral indicates protection of persons against access to hazardous parts inside enclosures or protection of equipment against ingress of solid foreign objects. The second numeral indicates protection of equipment against ingress of water. BS EN 62262:2002 specifies a system for classifying degrees of protection provided by enclosures against mechanical impact. Each characteristic group numeral represents an impact energy value. IK 07 represents impact energy in Joules of 2. More information on the IP classification code is given in IET Guidance Note 1: *Selection & Erection*.

20.6 Isolation and switching: emergency switching off

BS 7671:2018 (IET Wiring Regulations) recognizes four distinct types of isolation and switching operation:

(a) isolation;
(b) switching off for mechanical maintenance;
(c) emergency switching off; and
(d) functional switching.

Whilst Section 722 does not demand emergency switching off, it contains additional requirements stating that:

Where emergency switching off is required, such devices shall be capable of breaking the full load current of the relevant parts of the installation and disconnect all live conductors, including the neutral conductor.

537.3.3.3 It is worth noting that the Regulations state that plugs and socket-outlets shall not be provided for use as means for emergency switching off. Therefore, if emergency switching off is deemed necessary, the device will have to meet the requirements for emergency switching off contained in Regulations 465 and 537.3.3 of BS 7671.

20.7 Devices for fault protection by automatic disconnection of supply

722.531 Regulation 722.531.2.101 has been redrafted concerning RCD protection. The regulation now contains further requirements for both Type A and Type B RCDs to take account of DC fault current.

Regulation 722.531.2.101 contains the following requirements:

(a) each charging point shall be protected by its own RCD of at least Type A, having a rated residual operating current not exceeding 30 mA.
(b) for each charging point incorporating a socket-outlet or vehicle connector complying with the BS EN 62196 series, protective measures against DC fault current shall be taken, except where provided by the electric vehicle charging equipment. The appropriate measures, for each connection point, shall be as follows:
    (i) RCD Type B; or
    (ii) RCD Type A and appropriate equipment that provides disconnection of the supply in case of DC fault current above 6 mA.
(c) RCDs shall comply with one of the following standards: BS EN 61008-1, BS EN 61009-1, BS EN 60947-2 or BS EN 62423.

Note: Requirements for the selection and erection of RCDs in the case of supplies using DC vehicle connectors according to the BS EN 62196 series are under consideration. This means that the requirements are still being developed at European level.

Regulation 722.531.2.1.1 requires that RCDs shall disconnect all live conductors.

20.8 Other equipment: socket-outlets and connectors

722.55 Socket-outlets must be fit for purpose. They must be suitable for the load and for external influences such as protection against mechanical damage and ingress of water. Section 722 requires a degree of protection of at least IP44 where the equipment is installed outdoors. Portable socket-outlets such as extension leads shall not be used but tethered vehicle connectors are allowed. One socket-outlet or vehicle connector must supply only one electric vehicle.

Regulation 722.55.101.5 requires the lowest part of any socket-outlet to be placed at a height of 0.5 m to 1.5 m from the ground. However, the requirements of the relevant Building Regulations should also be adhered to in respect of socket-outlet heights.

Regulation 722.55.101.0.201.1 requires each AC charging point to incorporate:

(a) one socket-outlet complying with BS 1363-2 marked 'EV' on its rear. Except where there is no possibility of confusion, a label must be provided on the front face or adjacent to the socket-outlet or its enclosure stating: 'suitable for electric vehicle charging'; or

(b) one socket-outlet or connector complying with BS EN 60309-2 which is interlocked and classified to clause 6.1.5 of BS EN 60309-1 to prevent the socket contacts being live when accessible; or

(c) one socket-outlet or connector complying with BS EN 60309-2 which is part of an interlocked self-contained product complying with BS EN 60309-4 and classified to clauses 6.1.101 and 6.1.102 to prevent the socket contacts being live when accessible; or

Part 2 **(d)** for mode 3 charging only, a Type 1 vehicle connector complying with BS EN 62196-2, or a Type 2 socket-outlet (or vehicle connector) complying with BS EN 62196-2, or a Type 3 socket-outlet (or vehicle connector) complying with BS EN 62196-2.

Part 2 Regulation 722.55.101.4 requires that in EV charging modes 3 and 4, an electrical or mechanical system must be provided to prevent the plugging or unplugging of the plug unless the socket-outlet or the vehicle connector has been switched off from the supply.

Notes to charging modes:

Briefly, mode 1 and mode 2 charging utilizes standardized socket-outlets (mode 2 has an in-cable control box with personnel protection against electric shock and has a control pilot function). Mode 3 utilizes dedicated electric vehicle supply equipment and has a control pilot function. Mode 4 uses a DC connection of the EV to the AC supply network, utilizing an off-board charger where the control pilot function extends to equipment permanently connected to the AC supply.

20.9 Annex A to Section 722

20.9.1 Earth electrodes

542.2 BS 7671 recognizes a wide variety of types of earth electrode. Regulation 542.2.2 lists the types recognized, which include earth rods, earth plates and underground structural metalwork. The soil resistivity of the ground is probably the single most important factor in the determination of the type of earth electrode. It is therefore important when planning an earth electrode installation to ascertain the type of ground where the electrode is to be installed. For example, in some locations low soil resistivity is found to be concentrated in the topsoil layer, beneath which there may be rock or other impervious strata preventing the deep driving of rods, or a deep layer of high resistivity. In these circumstances, the installation of copper earth tapes, or pipes or plates, would be more likely to provide a satisfactory earth electrode resistance value.

Closely spaced buildings may sometimes make it difficult to find ground suitable for driving an earth electrode. Electrodes which employ suitable structural or other underground metalwork, or the metal reinforcement of concrete embedded in the ground, may then be of particular advantage. Where metal reinforcement of concrete is used, precautions should be taken to ensure there is electrical continuity between the reinforcing bars and the earth connections.

The first stage when planning an earth electrode installation is therefore to ascertain the type of ground, and, depending on the location, determine the type of electrode. The composition of the soil largely determines its resistivity. Earth resistivity is, however, essentially electrolytic in nature and is thus affected by the moisture content of the soil and by the chemical composition and concentration of salts dissolved in the contained water. Grain size and distribution, and closeness of packing, are also contributory factors, as they control the manner in which the moisture is held in the soil. Many of these factors vary locally and some vary seasonally.

BS 7430:2011+A1:2015 *Code of Practice for protective earthing of electrical installations* advises that where there is any option in locating earth electrodes, a site should be chosen in one of the following types of soil, in the order of preference given:

(a) wet marshy ground;
(b) clay, loamy soil, arable land, clayey soil or loam mixed with small quantities of sand;
(c) clay and loam mixed with varying proportions of sand, gravel, and stones; or
(d) damp and wet sand, peat.

Dry sand, gravel, chalk, limestone, whinstone, granite and any very stony ground should be avoided if possible, also all locations where virgin rock is very close to the surface.

BS 7430 also advises that a site should be chosen that is not naturally well-drained. A water-logged situation is not, however, essential, unless the soil is sand or gravel, as in general no advantage results from an increase in moisture content above about 15 % to 20 %. Care should be taken to avoid a site kept moist by water flowing over it (e.g. the bed of a stream), as the beneficial salts may be entirely removed from the soil in such situations.

Earth rods can only be as effective as the contact they make with the surrounding material. Where constructional work has taken place, or cut and fill or imported fill operations have been carried out, the resulting disturbance of the ground may alter site conditions. Therefore, earth rods should be driven into virgin ground and not into disturbed (backfilled) ground. Where disturbed ground cannot be avoided, deeper driving of the electrode may be necessary to reach layers of reasonable resistivity, and also to reach stable ground, such that the value of the earth electrode resistance remains stable if the top layers of the ground dry out.

Earth electrode contact resistance may be improved by soil treatment or replacement in special or difficult locations. However, migration and leaching of applied chemicals over a period of time reduces the efficiency of soil treatment. As a result, the installation would require constant monitoring and the additives would need to be replaced. Therefore this is only a temporary solution for temporary installations. Ecological considerations should be considered before such treatment is commenced and any deleterious effect upon electrode material has to be taken into account. As a more permanent solution, the soil immediately around the electrode could be replaced with a lower resistivity material. Such materials include bentonite, which is a clay-based material formed by the decomposition of volcanic ash, or a conductive concrete or cement made with graded granular carbonaceous aggregate in place of the conventional sand or aggregate.

BS 7430 contains examples of soil resistivity in ohms per metre for various types of soil. It is therefore possible to predict the resistance to earth of a vertical rod, for example, or a plate electrode or single strip electrode, by using this information in a formula and carrying out the appropriate calculation.

From this information, calculations can be carried out to predict the resistance to earth of the electrode and therefore the number and size of electrodes required to achieve a satisfactorily low earth electrode resistance.

Calculations on earth electrodes

Where it is required to predict the resistance to earth of a vertical rod, the following equation can be used:

$$R = \frac{\rho}{2\pi L}\left[\log_e\left(\frac{8L}{d}\right) - 1\right] (\Omega)$$

where:

L = the length of the electrode in metres
d = the diameter of the electrode in metres
ρ = the resistivity of the soil in ohm metres (Ωm)

Example 1:

L = 1.8 m long rod
d = 15 mm diameter rod
ρ = 10 for clay type soil from BS 7430

Therefore, calculating the resistance

$$R = \frac{10}{2\pi \times 1.8} \times \left[\log_e\left(\frac{8 \times 1.8}{0.015}\right) - 1\right] = 5.18 \ \Omega$$

Now compare this with an electrode in porous sandstone.

Example 2:

L = 1.8 m long rod
d = 15 mm diameter rod
ρ = 100 for porous sandstone from BS 7430

$$R = \frac{100}{2\pi \times 1.8} \times \left[\log_e\left(\frac{8 \times 1.8}{0.015}\right) - 1\right] = 51.87 \ \Omega$$

If we connect two rods together, we can now reduce the resistance of the earth electrode as follows:

Example 3:

$$R = \frac{100}{2\pi \times 3.6} \times \left[\log_e\left(\frac{8 \times 3.6}{0.015}\right) - 1\right] = 28.99 \ \Omega$$

Calculating the resistance of a plate electrode:

$$R = \frac{\rho}{4}\sqrt{\left(\frac{\pi}{2A}\right)}$$

where:

ρ is the resistivity of the soil in Ωm
A is the area of the plate in m^2

Example 4:

Therefore, assuming ρ = 10, and A = 1

$$R = \frac{10}{4}\sqrt{\left(\frac{\pi}{2 \times 1}\right)} = 3.13 \ \Omega$$

Now compare this with an electrode in porous sandstone:

Example 5:

ρ = 100 for porous sandstone from BS 7430

$$R = \frac{100}{4}\sqrt{\left(\frac{\pi}{2 \times 1}\right)} = 31.33 \ \Omega$$

Calculating the resistance to earth of a single strip electrode run in a straight line:

$$R = \frac{\rho}{2\pi L}\left[\log_e\left(\frac{2L^2}{wh}\right) - 1\right]$$

where:

L is the length of the strip conductor (m)
h is the buried depth of the electrode (m)
w is the width of the strip (m)
ρ is the resistivity of the soil in Ωm

Example 6:

L = 4 m long
h = 1 m
w = 30 mm
ρ = 10 for clay type soil from BS 7430

$$R = \frac{10}{2 \times \pi \times 4}\left[\log_e\left(\frac{2 \times 4^2}{0.03 \times 1}\right) - 1\right] = 2.38 \; \Omega$$

20.9.2 Earthing conductors

542.3.1 An earthing conductor, which is defined in BS 7671 as a protective conductor connecting the main earthing terminal of an installation to an earth electrode or other means of earthing, must be adequately sized, particularly where buried partly in the ground. It must be of suitable material and adequately protected against corrosion and mechanical damage.

The size of an earthing conductor is determined in basically the same way as a circuit protective conductor, except that Table 54.1 of BS 7671 must be applied to any buried earthing conductor. For a PME supply, it should be no smaller than the main bonding conductors.

Onshore units of electrical shore connections for inland navigation vessels

21

Section 730 is a new set of regulations included in Part 7 of the 18th Edition.

21.1 Introduction

Sect 730 Section 730 applies to onshore installations that are dedicated to the supply of inland navigation vessels for commercial and administrative purposes, berthed in ports and berths.

Most, if not all, of the measures used to reduce the risks in marinas apply equally to electrical shore connections for inland navigation vessels. One of the major differences between supplies to vessels in a typical UK marina and electrical shore connections for inland navigation vessels is the size of the supply needed. For example, vessels used on inland waterways in Europe can weigh up to 10,000 tonnes. These vessels are considerably larger than the average vessel used in a marina, which are generally small recreational craft, up to 24 m long.

Generally, socket-outlets with a rating of 16 A will be provided for each craft in a marina. Many of the risks associated with electrical installations in marinas, such as the presence of water, movement of structures and harsh environmental conditions, are similar for electrical shore connections for inland navigation vessels. This chapter summarizes some of the key requirements of Section 730.

21.2 Supplies

730.313 Section 730 requires that the nominal supply voltage shall be 400 V 3-phase AC 50 Hz.

Note: Where the supply system is PME, Regulation 9(4) of the ESQCR 2002 prohibits the connection of the neutral to the metalwork of any caravan or boat in the UK.

21.3 Galvanic separation

The immersion of metal components of a vessel in water, particularly in salt water, provides the natural mechanism of galvanic corrosion. Where there are dissimilar metals on the electrochemical series in proximity, the detrimental effect of galvanic couples can be exacerbated. For this reason, small vessels, recreational craft, houseboats, ships and many immersed metal structures are provided with sacrificial anodes (zinc for salt water), to which the more valuable/essential immersed metal parts, such as propellers, shafts, hull fittings and fixings, are electrically bonded. The sacrificial anode(s) preferentially deplete as a consequence of providing galvanic corrosion protection to such immersed parts.

Section 730 recognizes that there is an additional risk of electrolytic corrosion resulting from circulating galvanic currents in the protective conductor from the shore supply to a vessel.

There have also been reports of increased rate of depletion of the sacrificial anodes of vessels that are connected on a longer-term basis to shore supplies. This is believed by some observers to be associated with the connection of the vessel's protective earth terminal (to which immersed components and sacrificial anodes are bonded) to the shore-supply earth in an inland waterway or marina.

730.313.1.102 Section 730 recognizes the use of an isolating transformer to prevent galvanic currents circulating between the hull of the vessel and the metallic parts on the shore side. Where a fixed onshore isolating transformer is used to prevent galvanic currents circulating between the hull of the vessel and metallic parts on the shore side, equipment complying with BS EN 61558-2-4 shall be used.

21.4 Protection against electric shock

730.410.3.5 As would be expected, the protective measures of obstacles, placing out of reach, non-
730.410.3.6 conducting location and earth-free local equipotential bonding shall not be used. These measures are not for general application. They are only for application in installations controlled or supervised by skilled or instructed persons.

21.5 Operational conditions and external influences

Any wiring system or item of equipment selected and installed must be suitable for its location and able to operate satisfactorily during its working life. In ports and berths, consideration must also be given to the possible presence of corrosive or polluting substances.

Regulation 730.512.2.101 requires that equipment shall be selected with a minimum degree of protection of IP44.

21.6 Types of wiring system

730.521 Cables must be selected and installed so that mechanical damage due to tidal and other movement of floating structures is prevented.

Regulation 730.521.101.1 recognizes that the following wiring systems and cables are suitable for distribution circuits in berths and ports:

(a) underground cables;

(b) overhead cables;

(c) cables with copper conductors and thermoplastic or elastomeric insulation installed within an appropriate cable management system, taking into account external influences such as movement, impact, corrosion and ambient temperature;

(d) mineral-insulated cables with a thermoplastic protective covering;

(e) cables with armouring and serving of thermoplastic or elastomeric material; and

(f) other cables and materials that are at least as suitable as those listed above.

Regulation 730.521.101.2 recognizes that the following wiring systems and cables are suitable for distribution circuits on floating landing stages:

(a) cables with copper conductors and thermoplastic or elastomeric insulation, installed within an appropriate cable management system taking into account external influences such as movement, impact, corrosion and ambient temperature; and

(b) armoured cables with a thermoplastic or elastomeric covering.

Other cables and materials that are at least as suitable as those listed under (a) or (b) may be used.

Regulation 730.521.101.3.1 requires that underground distribution cables shall, unless provided with additional mechanical protection, be buried at a sufficient depth to avoid being damaged, for example, by vehicle movement. Overhead cables shall not be used over waterways (Regulation 730.521.101.3.3). Where overhead conductors are used, they must be insulated. Poles and other supports for overhead wiring must be located or protected so that they are unlikely to be damaged by any foreseeable vehicle movement.

Overhead conductors shall be at a height above ground of not less than 6 m in all areas subject to vehicle movement and 3.5 m in all other areas.

21.7 RCD protection

Regulation 730.531.3 states particular requirements concerning RCD protection:

730.531.3 **(a)** socket-outlets with a rated current up to 63 A shall be individually protected by an RCD providing additional protection in accordance with Regulation 415.1, having a rated residual operating current not exceeding 30 mA.
(b) the RCD selected shall disconnect all live conductors, i.e. line and neutral.
(c) socket-outlets with a rated current exceeding 63 A shall be individually protected by an RCD having a rated residual operating current not exceeding 300 mA. The RCD selected shall disconnect all live conductors, i.e. line and neutral.

Note: The purpose of these RCDs is to protect the shore supply and the flexible cable. They are not intended to provide protection for onboard circuits, which are outside the scope of Section 730.

21.8 Overcurrent protection

730.533 Similar to the requirements in marinas, socket-outlets shall be individually protected by an overcurrent protective device.

21.9 Isolation

730.537.2.1 Similar to the requirements in marinas, at least one means of isolation shall be installed for each distribution board. This device shall disconnect all live conductors.

21.10 Requirements for socket-outlets

730.55.1 Regulation group 730.55.1 sets out the following requirements for socket-outlets:

(a) socket-outlets shall comply with BS EN 60309-1 and BS EN 60309-4 and socket-outlets with a current rating up to and including 125 A shall comply with EN 60309-2;

(b) where interchangeability is not required, socket-outlets shall comply with BS EN 60309-1 and BS EN 60309-4 and need not comply with BS EN 60309-2;

(c) socket-outlets shall be located as close as practicable to the berth to be supplied;

(d) no more than four socket-outlets shall be grouped together in any one enclosure;

(e) each socket-outlet shall supply only one vessel;

730.553.13 **(f)** socket-outlets shall be placed at a height of not less than 1 m above the highest water level;

(g) in the case of floating pontoons or walkways only, this height may be reduced to 0.3 m above the highest water level, provided that appropriate additional measures are taken to protect against the effects of splashing; and

(h) socket-outlets shall be placed in an enclosure in accordance with BS EN 5869-2.

Note: Section 730 contains an Annex giving examples of methods of obtaining supply.

Index

W

X

Z